Springer Tracts in Modern Physics 118

Springer Tracts in Modern Physics

---

* denotes a volume which contains a Classified Index starting from Volume 36

Karl-Heinz Robrock

Mechanical Relaxation

of Interstitials in Irradiated Metals

With 67 Figures

Springer-Verlag Berlin Heidelberg GmbH

Dr. Karl-Heinz Robrock

Institut für Festkörperforschung
KFA Forschungszentrum Jülich GmbH, Postfach 1913
D-5170 Jülich, Fed. Rep. of Germany

Manuscripts for publication should be addressed to:

Gerhard Höhler

Institut für Theoretische Kernphysik der Universität Karlsruhe
Postfach 6980, D-7500 Karlsruhe 1, Fed. Rep. of Germany

*Proofs and all correspondence concerning papers in the process of publication
should be addressed to:*

Ernst A. Niekisch

Haubourdinstraße 6, D-5170 Jülich 1, Fed. Rep. of Germany

ISBN 978-3-662-15045-0   ISBN 978-3-540-46156-2 (eBook)
DOI 10.1007/978-3-540-46156-2

Library of Congress Cataloging-in-Publication Data. Robrock, Karl-Heinz, 1943–. Mechanical relaxation of interstitials in irradiated metals / Karl-Heinz Robrock. p. cm. – (Springer tracts in modern physics; 118) Bibliography: p. Includes index. 1. Metals – Effect of radiation on. 2. Stress relaxation. 3. Crystallography. I. Title. II. Series. QC1.S797 vol. 118 [TA418.6] 539 s–dc20   [620.1'628]   89-11331

© Springer-Verlag Berlin Heidelberg 1990

Originally published by Springer-Verlag Berlin Heidelberg New York in 1990.

Softcover reprint of the hardcover 1st edition 1990

2157/3150-543210 – Printed on acid-free paper

Preface

Mechanical relaxation studies serve to characterize the anelastic response of crystal defects to elastic stresses or strains. From the strength of the response, the magnitude and symmetry of the elastic distortions introduced by the defects are revealed, and from the characteristic response times information on the mobilities of the crystal defects is obtained. In this manner mechanical relaxation techniques have been employed to investigate the whole spectrum of lattice defects such as point defects, dislocations and grain boundaries. The subject is covered in two excellent textbooks, one by Nowick and Berry, the other by De Batist. Since the appearance of these two monographs in the early seventies, a large body of new experimental data and new experimental developments have arrived, particularly in the field of radiation-induced interstitial atoms in metals. The present book is meant to serve as an introduction into and an up-to-date review on this subject. In its theoretical background it is closely related to two theoretically oriented monographs in the STMP-series, namely volumes 81 and 87 on "Point Defects in Metals". For this reason the theoretical concepts are kept at an introductory level. The experimental techniques and results are presented in greater depth and detail. A critical assessment of the latest data relevant to the structural and dynamic properties of radiation-induced interstitial atoms provides a perspective on the state of the art.

Doing the research work and writing this book would not have been possible without the continued support and the guidance provided by W. Schilling, and the cooperation with C. Börner, H. G. Bohn, H. Jacques, G. Kollers, P. Okamoto, L. E. Rehn and V. Spiric, to all of whom I wish to extend my sincerest thanks. I gratefully acknowledge the stimulating discussions with P. H. Dederichs, P. Ehrhart and H. R. Schober. I wish to thank Mrs. M. Garcia for typing the manuscript with great competence, and for her special patience with its revisions.

Jülich, December 1989 *K.-H. Robrock*

Contents

1. Introduction

The first and most famous example of an interstitial-related anelastic relaxation effect was introduced by J. L. Snoek about 50 years ago. He explained the occurrence of internal friction peaks in α iron short above room temperature in terms of a simple model involving jumps of carbon or nitrogen atoms between neighboring interstitial sites. Due to their atomic mobility and their non-cubic site symmetry the carbon or nitrogen interstitial atoms are able to respond to external stresses in a manner that results in a characteristic deviation of the host material from its ideal elastic behavior, e. g., a reversible aftereffect or a mechanical damping occur. Both effects can be observed experimentally, and provide a sensitive and selective tool to study the structural and diffusional properties of the respective interstitial atoms.

In full analogy to this classical Snoek effect one can expect that interstitial atoms created athermally by irradiation with suitable high-energy particles give rise to corresponding anelastic relaxation effects. However, in contrast to the thermodynamically stable solid interstitial solution in the case of the Snoek effect, irradiation-induced interstitial atoms are basically unstable and prone to reactions or disintegration. In many cases they are, however, stable below a specific threshold temperature, and common cryogenic temperatures are sufficient to stabilize them and allow them to be studied. Although low temperature irradiations and measurements greatly increase the experimental effort and complexity, the underlying mechanisms are essentially the same as in the case of the original Snoek effect.

As pointed out by *Zener* [168] many other sources than interstitial atoms can account for the observation of anelastic relaxation effects. Pairs of solute atoms, dislocations, grain boundaries and thermal currents constitute a collection of possible candidates, which led Zener to propose a relaxation spectrum covering 20 decades in frequency! It is clear that such a wide range of frequencies cannot be covered experimentally.

Yet there is a unique approach which allows one to cover effectively ten and more decades on a frequency scale, namely in those cases where the atomic mobility is temperature-dependent, e. g., in the case of thermally activated processes. Depending on the details of the activation parameters, a small variation of the measuring temperature may result in a rather large variation of the underlying relaxation time and thus spectral range. As in the case of the original Snoek effect, anelastic relaxation spectra are most often acquired and shown as relaxation strength versus temperature rather than frequency.

The larger experimental effort accompanying irradiation-related anelastic studies is largely counterbalanced by a beneficial action of the irradiation, namely the suppression of dislocation and grain boundary-related relaxation effects. As many examples will show, this is a very welcome improvement in sensitivity with respect to the comparatively small relaxation effects due to irradiation-induced interstitial atoms.

This paper is arranged in the following manner. In the second, theoretically oriented chapter the dia- and paraelastic phenomena are outlined, and atomistic models of interstitial atoms are discussed. The third section contains a description of experimental techniques and devices. Chapter 4 covers the experimental results obtained from pure metals, Chapter 5 those from dilute alloys. Chapter 6 finally deals with the phenomenon of radiation-induced segregation, which is shown to be intimately related to the action of the interstitial atoms described in the previous sections.

2. Theoretical Background

2.1 Dipole Force Tensor and Elastic Polarization

2.1.1 Structures of Interstitials

Experimental information on interstitial structures come mainly from scattering experiments and from the mechanical polarization experiments which will be discussed extensively in this paper. These experimental investigations were paralleled by theoretical work, analytical and to a larger extent computer simulations. The results may be summarized as follows [1, 2]:

The configurations of self-interstitials in cubic metals are shown in Fig. 2.1. They consist of doubly occupied lattice sites, i.e., two atoms share a site together in the form of a so-called dumbbell. The orientation of the dumbbell-axis is specific for the lattice type: it is <100> in the fcc and <110> in the bcc structure. In the fcc lattice there are three equivalent <100> orientations whereas in the bcc structure there are six equivalent <110> orientations.

The dumbbell distance, d, is smaller than the regular atomic equilibrium distance R_0, typically $d/R_0 = 0.80$ in the fcc lattice [3]. The atoms close to the dumbbell are pushed outwards from their ideal lattice sites, but still have distances smaller than R_0, i.e., the whole atomic arrangement is highly compressed in the vicinity of the dumbbell with respect to regular atomic separations. Many other configurations have been conceived for self-interstitials in cubic metals, for instance with the extra atom occupying the respective octahedral and tetrahedral interstitial positions, or dumbbells with <110> (fcc) or <111> (bcc) axis orientation. However, as discussed extensively in a recent review [2], they have all been excluded on the basis of theoretical considerations as well as experimental observations.

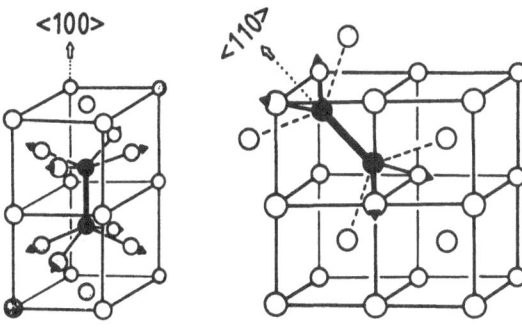

Fig. 2.1. Configurations of self-interstitials in fcc metals (e.g. Al, Cu or Ni), and bcc metals (e.g., Fe or Mo)

3

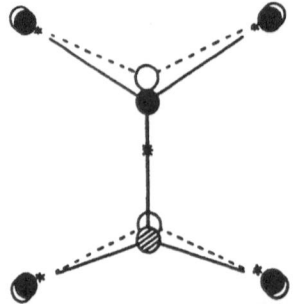
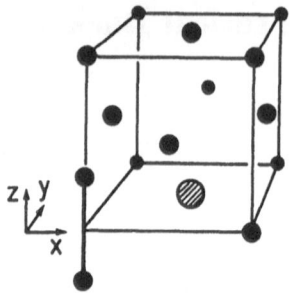

Fig. 2.2. Structure of the mixed dumbbell as compared to a "regular" dumbbell according to [3]. •: atomic positions for the mixed dumbbell: o: atomic positions for the "regular" dumbbell: *: positions in the regular lattice: ◑: position of the solute atom

Fig. 2.3. Complex consisting of a "regular" dumbbell bound to an oversized solute atom, ◑, on a nearest neighbor position

Detailed atomistic interstitial models have also been proposed for dilute substitutional alloys. Here, the solute atoms may enter the interstitial configurations either if a replacement collision sequence terminates at the solute atom, or otherwise if pure solvent self-interstitials are able to migrate during or after irradiation and to react with the solute atoms.

In terms of a model proposed by *Dederichs* et al. [3], the resulting interstitial configuration depends on whether a given solute atom is undersized or oversized with respect to the solvent atom.

The mixed dumbbell shown in Fig. 2.2 is one example of how undersized solute atoms may be positioned on interstitial sites, namely by replacing one host atom of the dumbbell. The originally substitutional foreign atom thus becomes an interstitial atom. On the other hand, as shown in Fig. 2.3, oversized solute atoms remain basically on the regular site and bind a pure solvent atom dumbbell on a nearest neighbor site. Further details on this subject are discussed in Chap. 5.

2.1.2 Dipole Tensors

When a lattice site becomes doubly occupied as with the self-interstitial dumbbell the surrounding atoms have to be shifted outwards in order to accommodate the extra atomic volume. In this manner a displacement field is set up in the crystal. It falls off with distance, r, from the defect center as $s(r) \propto r^{-2}$ for large distances $r \gg d$ [1]. The strength of the displacement field is proportional to the first moment $\boldsymbol{P}$, of a distribution of forces, $\boldsymbol{K}^m$, acting on atoms labelled m, at lattice vectors, $\boldsymbol{X}^m$:

$$P_{ij} = \sum_m X_i^m K_j^m . \tag{2.1}$$

The forces, $\boldsymbol{K}^m$, are indicated by arrows in Fig. 2.1.

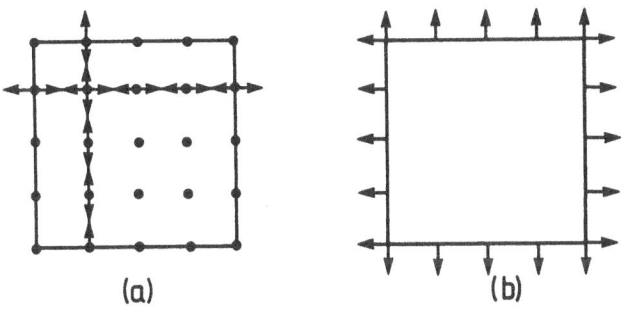

Fig. 2.4. Schematic arrangement of simple double-forces. The forces are cancelled out in the interior of the body (**a**) and only surface forces remain (**b**). This force arrangement is equivalent to normal stresses at the surfaces of the crystal

They represent the repulsive action of the defect on its neighbor atoms. The relaxation volume of an interstitial is defined as

$$V^{\text{rel}} = \text{Tr}(\underline{P})/3K \tag{2.2}$$

where K is the elastic bulk modulus, V^{rel} is the total volume change of a crystal of finite size, if an additional atom is placed on an interstitial site. The volume change is independent of the position of the point defect in the crystal [1]. Therefore the volume change of a material containing a distribution of many non-interacting defects of number, n, is given by

$$\frac{\Delta V}{V} = \frac{n}{N}\frac{V^{\text{rel}}}{\Omega} = c\frac{V^{\text{rel}}}{\Omega} = \varrho\frac{\text{Tr}(\underline{P})}{3K} \tag{2.3}$$

where N is the number of lattice atoms, Ω the atomic volume, c the atomic fraction and ϱ the volume density of defects. The situation is depicted in Fig. 2.4. The defects are symbolized by the respective double forces (Fig. 2.4a). Since the action of the mutually opposing forces is cancelled out in the interior, the crystal suffers an expansion as due to surface forces, which are equivalent to external stresses (Fig. 2.4b). For the stress component σ_{xx} one finds for instance

$$\sigma_{xx} = 2K_x \cdot \varrho \cdot a_x = \varrho P_{xx}$$

where ϱa_x is the number of forces per unit surface area coming from defects at a depth of thickness a_x. If we identify a_x as the atomic nearest neighbor distance, R_0, the product $2a_x K_x$ corresponds to the component of the dipole tensor, P_{xx}, of the defect.

Anisotropic defects, e. g, dumbbells as self-interstitials, may be oriented in different lattice directions as explained in Sect. 2.1.1. If we denote these directions by an index (ν), and add up their contributions we obtain

$$\underline{\sigma} = \sum_{\nu} \varrho^{(\nu)}\underline{P}^{(\nu)} \tag{2.4}$$

5

which is in accordance with the exact derivation [1]. If the crystal is allowed to relax under these stresses, the resulting strain is given by [12]

$$\underline{\varepsilon} = \sum_{\nu} c^{(\nu)} \underline{\lambda}^{(\nu)} . \tag{2.5}$$

From (2.4) and (2.5) it follows that

$$\underline{P}^{(\nu)} = \frac{\partial \underline{\sigma}}{\partial \varrho^{(\nu)}} \quad \text{and} \quad \underline{\lambda}^{(\nu)} = \frac{\partial \underline{\varepsilon}}{\partial c^{(\nu)}} \tag{2.6}$$

i.e. the dipole tensor $\underline{P}^{(\nu)}$ is the contribution of one defect per unit volume of orientation (ν) to the dilatational stresses, and $\underline{\lambda}^{(\nu)}/N$ the strain produced per defect of orientation (ν). The relation between $\underline{P}$ and $\underline{\lambda}$ for any orientation (ν) follows from Hooke's law:

$$\underline{\sigma} = \underline{M} \underline{\varepsilon}; \quad \underline{P} = \Omega \underline{M} \underline{\lambda} \tag{2.7}$$

where $\underline{M}$ is the fourth rank tensor describing the elastic moduli of the matrix.

For the <100> and <110> dumbbell configuration the structure of the dipole tensor or $\underline{\lambda}$ tensor may be derived from simple symmetry considerations. If a rectangular coordinate system is used with its axis parallel to the cubic <100> axes, then

$$\underline{P}^{(1)}_{<100>} = \begin{pmatrix} P_1 & 0 & 0 \\ 0 & P_2 & 0 \\ 0 & 0 & P_2 \end{pmatrix} \quad \text{and}$$

$$\underline{P}^{(1)}_{<110>} = \frac{1}{2} \begin{pmatrix} P_1 + P_2 & P_1 - P_2 & 0 \\ P_1 - P_2 & P_1 + P_2 & 0 \\ 0 & 0 & 2P_3 \end{pmatrix} \tag{2.8}$$

where index (1) pertains to the orientation of the principle axes e_1 of the dipole tensors as outlined in Table 2.1, and indexes <100> and <110> distinguish between dumbbells with <100> and <110> axis orientation, respectively. The simply indexed quantities P_α are the eigenvalues of the dipole tensors. $\underline{P}_{<100>}$ possesses a <100> tetragonal symmetry ($P_2 = P_3$) and $\underline{P}_{<110>}$ a <110> orthorhombic symmetry. The dipole force tensor for the other orientations (ν) may be derived from the preceding forms by appropriate coordinate transformations. In the tetragonal case this is reduced to simply permuting the components.

More details are given in Table 2.1. The first column lists the principle axes of the dipole tensors of different symmetries, together with the eigenvalues P_α assigned to them. The second column lists the dipole tensors for orientation (1) expressed in terms of the eigenvalues P_α. The third column lists a possible choice of orientation vectors (e_1) of equivalent orientations. The quantities ν_{ij} in this row are discussed later. In addition to the <100>

Table 2.1.*

Principle axes e_i and the respective P-components	Dipole-tensor in cubic axes	Equivalent orientation of axes e_1 and jump frequencies ν_{ij}	Polarizability α[2] and related jump rate $1/\tau'$	Polarizability α[4] and related jump rate $1/\tau$
Tetragonal:				
e_1: 100, P_1 e_2: 010, P_2 e_3: 001, P_2	$\begin{bmatrix} P_1 & 0 & 0 \\ & P_2 & 0 \\ & & P_2 \end{bmatrix}$	(1): 100 (2): 010 (3): 001 $\nu_{12} = \nu_{13}$	$\frac{1}{3kT}(P_1 - P_2)^2$ $3\nu_{12}$	0 —
Trigonal:				
e_1: 111, P_1 e_2: 1$\bar{1}$0, P_2 e_3: 11$\bar{2}$, P_2	$\frac{1}{3}\begin{bmatrix} P_1+2P_2 & P_1-P_2 & P_1-P_2 \\ P_1-P_2 & P_1+2P_2 & P_1-P_2 \\ P_1-P_2 & P_1-P_2 & P_1+2P_2 \end{bmatrix}$	(1): 111 (2): $\bar{1}$11 (3): 1$\bar{1}$1 (4): 11$\bar{1}$ $\nu_{12} = \nu_{13} = \nu_{14}$	0 —	$\frac{2}{9kT}(P_1 - P_2)^2$ $4\nu_{12}$
110 Orthorhombic:				
e_1: 110, P_1 e_2: 1$\bar{1}$0, P_2 e_3: 001, P_3	$\frac{1}{2}\begin{bmatrix} P_1+P_2 & P_1-P_2 & 0 \\ P_1-P_2 & P_1+P_2 & 0 \\ & & 2P_3 \end{bmatrix}$	(1): 110 (2): 1$\bar{1}$0 (3): 011 (4): 01$\bar{1}$ (5): 101 (6): 10$\bar{1}$ $\nu_{12} \neq \nu_{13}$ $\nu_{13} = \nu_{14} = \nu_{15} = \nu_{16}$	$\frac{1}{12kT}(P_1 + P_2 - 2P_3)^2$ $6\nu_{13}$	$\frac{1}{6kT}(P_1 - P_2)^2$ $2\nu_{12} + 4\nu_{13}$

* Vector brackets, [ ], of e_i are omitted for clarity

7

tetragonal (first line) and <110> orthorhombic case (third line) also the case of trigonal symmetry is given.

2.1.3 Paraelastic Relaxation Strengths

If point defects with dipole tensors $\underline{P}^{(\nu)}$ are subjected to an external strain $\underline{\varepsilon}^{\text{ext}}$, work is done by the defect forces K^m since the atoms are displaced by s^m from their equilibrium sites. For a defect with orientation (ν) the total work is:

$$W^{(\nu)} = -\sum_m K^{(\nu)m} \cdot s^m .\tag{2.9}$$

Under a homogeneous deformation

$$s_i^m = \sum_j \varepsilon_{ij}^{\text{ext}} X_j^m \tag{2.10}$$

so that

$$W^{(\nu)} = -\sum_{i,j} P_{ij}^{(\nu)} \varepsilon_{ij}^{\text{ext}} = -\Omega \sum_{ij} \lambda_{ij}^{(\nu)} \sigma_{ij}^{\text{ext}} .\tag{2.11}$$

This means that the energies of the various orientations can be shifted with respect to each other by the application of an external field, $\underline{\varepsilon}^{\text{ext}}$ or $\underline{\sigma}^{\text{ext}}$.

The equilibrium distribution of the point defect over the possible orientations is given by a Boltzmann distribution [12]

$$\varrho^{(\nu)} = \varrho_0 \frac{\exp(-W^{(\nu)}/kT)}{\sum_\mu \exp(-W^{(\mu)}/kT)} \tag{2.12}$$

where $\varrho_0 = \sum_\mu \varrho^{(\mu)}$ is the total concentration of defects. If transitions among the different orientations are possible, a redistribution of the defect orientations will occur and thus a change of the lattice expansion (2.3) as shown in Fig. 2.5. In order to arrive at simple formulae we write the dipole tensor as [1]

$$\underline{P}^{(\nu)} = \underline{P}_0 + \delta\underline{P}^{(\nu)} ,$$

where

$$\underline{P}_0 = \tfrac{1}{3}\text{Tr}(\underline{P}^{(\nu)}) \quad \text{and} \quad \sum_\nu \delta\underline{P}^{(\nu)} = 0 .\tag{2.13}$$

The differences in $W^{(\mu)}$ are usually much smaller than kT so that one can expand (2.12) in powers of $\underline{\varepsilon}$. If one confines the calculation to terms linear in $\underline{\varepsilon}$ one obtains from (2.4) and (2.12)

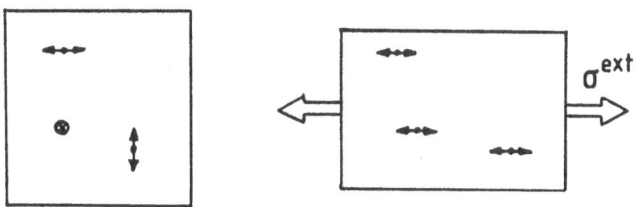

Fig. 2.5. Transition from the equilibrium distribution under zero stress (*left side*) to the alignment of double forces (*right side*) due to the application of an external stress σ^{ext} (schematic)

$$\Delta\underline{\sigma} = \underline{\sigma} - \varrho_0\underline{P}_0 = \frac{\varrho_0}{zkT} \sum_{\nu,k,l} \delta P_{ij}^{(\nu)} \delta P_{kl}^{(\nu)} \varepsilon_{kl}^{ext} . \tag{2.14}$$

where z is the number of possible orientations.

The change of the polarization stress, $\Delta\underline{\sigma}$, characterizes the response of the elastic dipoles to an external load. The stress which is necessary to produce a strain $\underline{\varepsilon}$, is smaller by an amount $\Delta\underline{\sigma}$ in a crystal containing the orientable defects as compared to a crystal free of defects. In turn, if the crystal with defects is subjected to an external stress, $\underline{\sigma}^{ext}$, the total strain of the crystal is larger by an amount

$$\Delta\underline{\varepsilon} = \underline{M}^{-1}\Delta\underline{\sigma} \tag{2.15}$$

as compared to a defect free crystal.

The polarizability, $\underline{\alpha}$, a fourth rank tensor is defined by

$$\Delta\underline{\sigma} = \varrho_0\underline{\alpha}\,\underline{\varepsilon}^{ext} . \tag{2.16}$$

The stress, $\Delta\underline{\sigma}$, or the polarizability, $\underline{\alpha}$, may alternatively be expressed in terms of a modulus change, $\Delta\underline{M}$, by the definition

$$\Delta\underline{\sigma} = -\Delta\underline{M} \cdot \underline{\varepsilon}^{ext} = \underline{M} \cdot \Delta\underline{\varepsilon} \tag{2.17}$$

i. e.

$$\Delta\underline{M} = -\varrho_0 \cdot \underline{\alpha} .$$

The modulus change, $\Delta\underline{M}$, has the character of a susceptibility. The change in stress, $\Delta\underline{\sigma}$, or strain, $\Delta\underline{\varepsilon}$, is referred to as paraelastic polarization by analogy to paramagnetic effects. It results from the "alignment" of permanent elastic dipoles with respect to the deformation axes of an elastically deformed crystal. As an illustration, we discuss the case of the <100> and <110> dumbbell in somewhat more detail.

For this purpose it is practical to make use of a special representation of the stresses and strains in terms of a set of six orthonormal basis tensors adapted to cubic symmetry [1, 12]. Figure 2.6 displays the deformations $\underline{\varepsilon}^{[\nu]}$ of a cube due to these basis tensors.

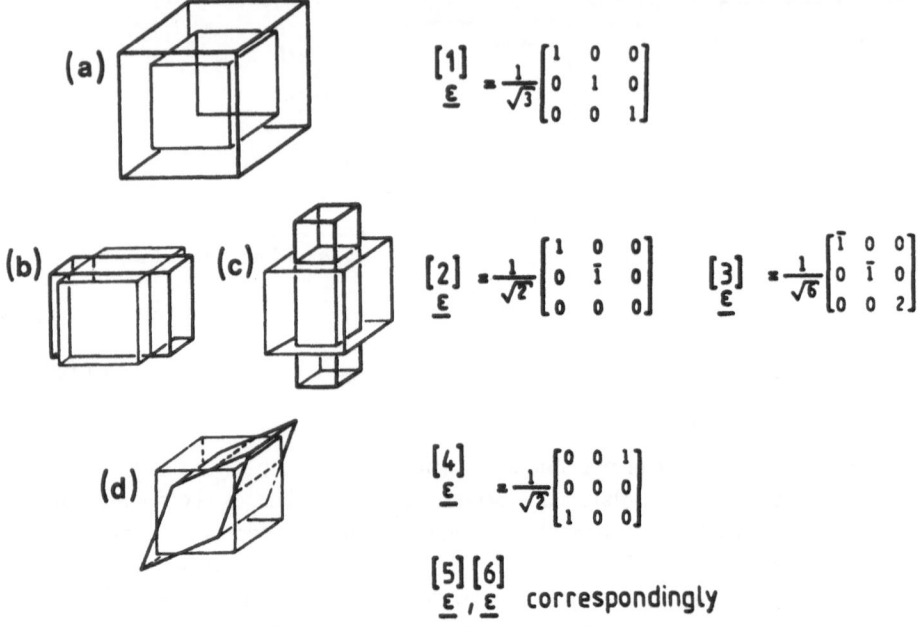

Fig. 2.6a–d. Deformation pattern of a unit cube due to the basis tensors adapted to cubic symmetry. Figure from [1]

The first mode $\overset{[1]}{\underline{\varepsilon}}$ corresponds to a homogeneous dilatation. Modes $\overset{[2]}{\underline{\varepsilon}}$ and $\overset{[3]}{\underline{\varepsilon}}$ are <110> type shear deformations. They result from a superposition of atomic displacements in <110> directions. Finally, mode $\overset{[4]}{\underline{\varepsilon}}$ represents one of three equivalent <100> shear modes ($\overset{[4]}{\underline{\varepsilon}}$ to $\overset{[6]}{\underline{\varepsilon}}$), which result from <100> type displacement in {100} planes. If the stresses, strains and elastic moduli are expressed in this basis, the stress-strain relations decouple and one obtains a set of three mutually independent scalar equations:

$$\overset{[\nu]}{\underline{\sigma}} = \overset{[\nu]}{C} \cdot \overset{[\nu]}{\underline{\varepsilon}} \quad \text{and} \quad \overset{[\nu]}{\underline{P}} = \Omega \overset{[\nu]}{C} \overset{[\nu]}{\underline{\lambda}} \tag{2.18}$$

where

$$\overset{[1]}{C} = C_{11} + 2C_{12} = 3K \tag{2.19a}$$

$$\overset{[2]}{C} = \overset{[3]}{C} = C_{11} - C_{12} = 2C' \tag{2.19b}$$

$$\overset{[4]}{C} = \overset{[5]}{C} = \overset{[6]}{C} = 2C_{44} = 2C . \tag{2.19c}$$

The $C_{\alpha\beta}$ are the elasticity moduli as given in Voigt's notation; K, C, and C' are commonly referred to as the compression modulus and the shear moduli, respectively.

The polarizabilities follow from (2.14) and (2.16):

$$\overset{[1]}{\alpha} = 0 \tag{2.20a}$$

$$\overset{[2]}{\alpha} = \frac{1}{6kT}[(P_{11} - P_{22})^2 + (P_{22} - P_{33})^2 + (P_{11} - P_{33})^2] = \overset{(3)}{\alpha} \tag{2.20b}$$

$$\overset{[4]}{\alpha} = \frac{2}{3kT}\left[P_{12}^2 + P_{13}^2 + P_{23}^2\right] = \overset{(5)}{\alpha} = \overset{(6)}{\alpha}. \tag{2.20c}$$

The P_{ij}'s can be used instead of the δP_{ij}'s, since the contribution from P_0 drops out. Also, the P_{ij}'s may be chosen for any defect orientation (ν). The change of the moduli $\overset{[\nu]}{C}$ is given by

$$\Delta \overset{[\nu]}{C} = -\varrho_0 \overset{[\nu]}{\alpha}. \tag{2.21}$$

Polarizabilities $\overset{[2]}{\alpha}$ and $\overset{[4]}{\alpha}$ are listed in Table 2.1 for several simple defect symmetries. The deformation patterns [1] to [6] allow one to forecast qualitatively the paraelastic response of a point defect with given symmetry by visual inspection. For instance, one can "see" directly that a homogeneous dilatation or compression $(\overset{[1]}{\varepsilon})$ cannot give rise to an orientational redistribution irrespective of the kind of dipole tensor, since no lattice direction is distinguished, i.e. $\overset{[1]}{\alpha} = 0$ as stated previously.

The <100> tetragonal dipole tensor, characteristic of a <100> dumbbell, is a particularly illustrative example to demonstrate the action of the shear deformations $\overset{[2]}{\varepsilon}$ to $\overset{[6]}{\varepsilon}$. Figure 2.7b shows a two-dimensional sketch of orientation (1) and (3) under the $\overset{[2]}{\varepsilon}$ mode. It is apparent that orientation (1) is energetically favored over orientation (3) since in the first case the largest tensile component of the dipole tensor is aligned with the dilatational component of the deformation, whereas the situation is the opposite in the second case. Therefore, as schematically indicated in Fig. 2.7a, the different orientations are repopulated according to (2.12) which results in a polarization stress

$$\Delta \overset{[2]}{\underline{\sigma}} = \frac{1}{3}\frac{\varrho_0}{kT}(P_1 - P_2)^2 \cdot \overset{[2]}{\underline{\varepsilon}}.$$

Under a $\overset{[4]}{\underline{\varepsilon}}$-, $\overset{[5]}{\underline{\varepsilon}}$-, $\overset{[6]}{\underline{\varepsilon}}$ deformation the two orientations (1) and (3) remain equivalent, i.e. no response of the <100> defect will occur, i.e.

$$\Delta \overset{[4]}{\underline{\sigma}} = \Delta \overset{[5]}{\underline{\sigma}} = \Delta \overset{[6]}{\underline{\sigma}} = \Delta \overset{[4]}{\underline{\varepsilon}} = \Delta \overset{[5]}{\underline{\varepsilon}} = \Delta \overset{[6]}{\underline{\varepsilon}} = 0.$$

We recognize that since $\overset{[4]}{\underline{\varepsilon}}$ to $\overset{[6]}{\underline{\varepsilon}}$ contain only off-diagonal terms, coupling can only occur if the dipole tensor possesses corresponding terms. For the <100>

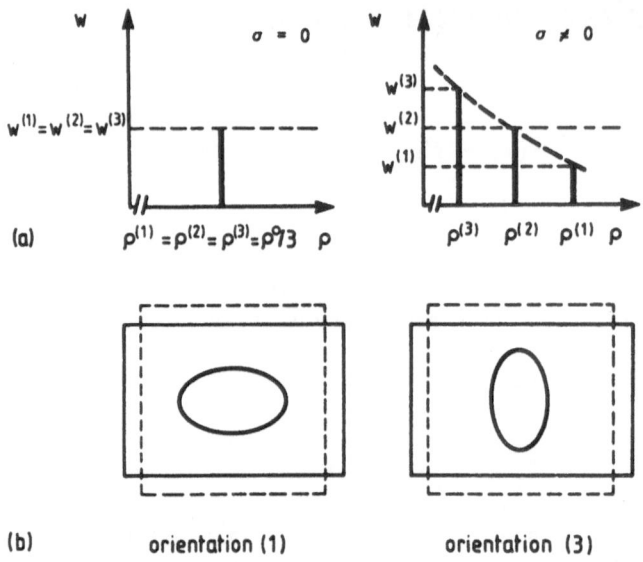

(b)　　　　orientation (1)　　　　orientation (3)

Fig. 2.7a,b. A two-dimensional sketch of the two orientations (1) and (3) of a <100> tetragonal dipole tensor under deformation mode $\overset{[2]}{\varepsilon}$. The dotted lines represent the undeformed cube in (b). (a) shows the corresponding energy levels $W^{(\nu)}$ and densities $\varrho^{(\nu)}$

tetragonal defect this is not the case, the energies do not change under a $\overset{[4]}{\varepsilon}$ to $\overset{[6]}{\varepsilon}$ type deformation, the polarization is zero.

In this manner, the response of point defects in cubic crystals to external deformations may be characterized by a simple rule: The orientational response to $\overset{[1]}{\varepsilon}$ is always zero, a response to $\overset{[2]}{\varepsilon}, \overset{[3]}{\varepsilon}$ occurs via the diagonal terms, and only if at least two of them are different from each other. A response to $\overset{[4]}{\varepsilon}$ to $\overset{[6]}{\varepsilon}$ occurs via the off-diagonal terms of the respective dipole tensor.

Another advantage of the preceding formulation is to be found in the fact that $\overset{[1]}{\varepsilon}$ to $\overset{[6]}{\varepsilon}$ are deformations which can be very easily realized experimentally. For instance, a $\overset{[4]}{\varepsilon}$ type shear occurs under torsion of a single crystal whose axis parallels a <100> lattice direction. The subject of orientation dependences will be discussed in more detail in Sect. 3.4.

2.1.4 Comparison with Diffuse Huang X-Ray Scattering

If a scattering experiment is carried out on a crystal containing interstitials, the scattering pattern is altered in a characteristic manner with respect to that of an ideal crystal. This is shown schematically in Fig. 2.8, where in the lower part the scattered intensity, I, is plotted as a function of the scattering vector

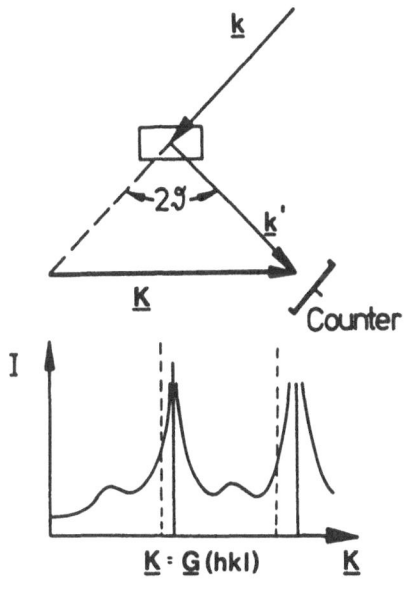

Fig. 2.8. Upper part, schematic arrangement of a scattering experiment: k and k' are the wave vectors of the incident and the scattered waves, respectively, and $K = k' - k$. The lower part shows a schematic diffraction pattern. The dotted lines represent two Bragg reflections of an ideal crystal. The solid lines represent intensities as observed from crystals containing defects exhibiting a shift of the position of $K = G$ and a diffuse background around the Bragg reflection

$K = k' - k$, where k' and k are the outgoing and incoming wave vectors, respectively, as indicated in the upper part of Fig. 2.8. For the ideal crystal one finds sharp intensity maxima, the so-called Bragg peaks, at positions $K = G$, the vectors of the reciprocal lattice. In a crystal containing point defects such as interstitials the changes which occur may be classified in three separate effects:

a) The location of the Bragg peaks is shifted by an amount, ΔG, indicating that the lattice has expanded by an average amount Δa. This lattice expansion is caused by the permanent stresses $\underline{\sigma} = \varrho_0 \underline{P}_0$ (2.14).
b) The intensity of the original Bragg peaks is reduced. This can be described in terms of a static Debye Waller factor, $\exp(-K^2 <s^2>)$ where s is the static atomic displacement [14].
c) Diffuse scattering intensity is observed between the Bragg peaks. Close to the Bragg reflection, this intensity varies as q^{-2} where $q = |G - K|$ and is called Huang scattering.

For cubic crystals the Huang scattering intensity is given by [14]:

$$I(K) = cN|f|^2 \left(\frac{G^{<hkl>}}{q^{<hkl>}} \right)^2$$

$$\times \frac{1}{\Omega^2} [\gamma^{(1)}(K)\pi^{(1)} + \gamma^{(2)}(K)\pi^{(2)} + \gamma^{(3)}(K)\pi^{(3)}] . \tag{2.22}$$

cN is the number, n, of defects in the crystal and f the atomic scattering factor. Particularly interesting in the present context are the quantities $\pi^{(i)}$, which are given by

13

$$\pi^{(1)} = \tfrac{1}{3}[P_{11} + P_{22} + P_{33}]^2 \tag{2.23a}$$

$$\pi^{(2)} = kT \overset{[2]}{\alpha} \tag{2.23b}$$

$$\pi^{(3)} = kT \overset{[4]}{\alpha} . \tag{2.23c}$$

The comparison of (2.20) with (2.23) reveals the correlation between the paraelastic polarizabilities and the Huang diffuse scattering intensities. Both are determined by the same combinations of the components of the double force tensor. The difference between (2.20) and (2.23) namely $\overset{[1]}{\alpha} = 0$ and $\pi^{(1)} \neq 0$ arises from the restriction of the paraelastic response to homogeneous deformations. Under inhomogeneous compressional deformations, $\overset{[1]}{\varepsilon}$, a polarizability is given by [15]

$$\overset{[1]}{\alpha}(\text{inhomog.}) = \frac{1}{3kT}[P_{11} + P_{22} + P_{33}]^2 = \frac{1}{kT}\pi^{(1)} .$$

This is the polarizability observed in diffusional relaxation experiments (Gorsky effect).

The coefficients $\gamma^{(1)}(K)$ in (2.22) are expressions composed of certain combinations of elastic moduli and scattering vectors K and q which we will not discuss in detail. The important point is that they can be made zero deliberately by a suitable selection of the Bragg reflection, G, and of the direction of the scattering vector, q [14]. For instance, for a reflection of type $G = [h00]$ the three components $\pi^{(i)}$ can be determined separately:

$$q \parallel [110] : I \propto (4\pi^{(1)} + 2\pi^{(2)} + 3\pi^{(3)}) \tag{2.24a}$$

$$q \parallel [110] : I \propto \pi^{(2)} \tag{2.24b}$$

$$q \parallel [001] : I \propto \pi^{(3)} . \tag{2.24c}$$

This procedure of separating the different $\pi^{(i)}$ from each other by a suitable choice of reflections and directions is paralleled in the separation of the different $\overset{[\nu]}{\alpha}$ by the choice of a suitable deformation, $\overset{[\nu]}{\varepsilon}$.

2.1.5 Diaelastic Polarization

In addition to the static displacements around a defect discussed in the previous section a second important modification of the crystal occurs in the neighborhood of a point defect. This is the change of the force constants or coupling parameters of the interatomic coupling forces [1, 11]. For a homogeneous distribution of many defects, this leads to a change of the elastic behavior of the crystal containing the defects. The effect turns out to be particularly pronounced for self-interstitials. It will be discussed in the following in terms of a simple spring model for the <100> dumbbell interstitial.

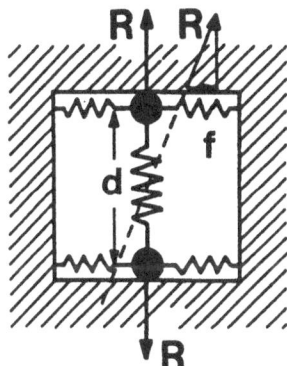

Fig. 2.9. Simple mechanical model of the dumbbell self-interstitial embedded in a rigid or elastic medium

Since an extra atom is accommodated interstitially, the atomic arrangement is locally highly compressed. If the interatomic force constants are visualized as springs connecting the atoms the spring between the dumbbell atoms and the springs to their respective neighboring atoms may be regarded as compressed. The situation can be modelled by a dumbbell which is embedded in a narrow rigid box, which represents the remainder of the crystal (Fig. 2.9). The dumbbell atoms are pushed in longitudinal directions against the box walls with a force $R = \partial V / \partial r|_{r=d}$ where V is the interatomic potential and d the dumbbell distance.

To simplify the model, the dumbbell atoms are connected to their respective neighbors, represented by the side walls of the box, by normal lattice springs with force constants f. If the dumbbell is twisted with respect to the box (fixed crystal) by an angle ψ as indicated by the dashed line, the atoms experience a tangential force component due to their displacement by an amount s

$$F_{\mathrm{t}} = R \sin \psi = R \frac{2s}{d}$$

in addition to the "normal" spring forces

$$F_{\mathrm{s}} = -2f \cdot s \ .$$

Thus the total restoring force, $F_{\mathrm{tot}} = F_{\mathrm{t}} + F_{\mathrm{s}}$, possesses an effective spring constant

$$f_{\mathrm{eff}} = 2 \left(f - \frac{R}{d} \right) \ . \tag{2.25}$$

In this manner, each self-interstitial causes a local reduction of the static restoring force of a crystal against certain deformation modes and thereby a marked modification of the elastic behavior of the crystal. The forces F_{t} may be regarded as induced forces, since they arise only if the dumbbell atoms are displaced, for instance by external shear forces. The first moment of these

15

forces comprises a second contribution to the dipole tensor, $\underline{P}^{ind}$, in addition to the permanent component $\underline{P}$ introduced previously

$$\underline{P}^{ind} = \sum_m X^m F_t^m .$$ (2.26)

The magnitude of its elements may be estimated as

$$|X| \cdot |F_t| = \frac{a}{2} \cdot R \cdot \frac{2s}{d} \approx R \cdot |X| \cdot \varepsilon \approx R \cdot \frac{a}{2} \cdot \varepsilon$$

where $|X| = a/2$ has been assumed as an order of magnitude value, and a homogeneous deformation $S = \varepsilon X$. In more general terms and taking different defect orientations (ν) into account, we obtain

$$\underline{P}^{(\nu)ind} = \underline{\alpha}_d^{(\nu)} \cdot \underline{\varepsilon} .$$ (2.27)

$\underline{\alpha}_d^{(\nu)}$ is the diaelastic polarizability. The index (ν) has been included here to express that interstitials with a different orientation (ν) may possess different induced dipole tensors because of their different orientation with respect to $\underline{\varepsilon}$.

If interstitials exhibit lower than cubic symmetry and occupy different equivalent orientations a proper average has to be taken in order to calculate the diaelastic polarizabilities. In addition to the springs between the dumbbell atoms also those between the dumbbell atoms and their nearest neighbors must be taken into account which was neglected in our simple example above. Thus

$$\underline{\alpha}_d = < \underline{\alpha}_d^{(\nu)} >_\nu .$$ (2.28)

The corresponding diaelastic modulus change is by analogy to (2.17) given by

$$\Delta \underline{M}_d = -\varrho^0 \cdot \underline{\alpha}_d .$$ (2.29)

The simulation of the crystal surrounding the dumbbell interstitial by a rigid box as above is another restriction which must be dropped for a more realistic model. This has been done for instance in the computer simulation by placing the dumbbell inside a crystallite of some thousand atoms, and by placing this crystallite into an elastic frame [2].

The reduced spring constants [3,7,8,9] manifest themselves in low frequency vibrational modes. Qualitatively the low frequency follows from the relation

$$\omega_R^2 = \frac{f_{eff}}{M_{eff}}$$ (2.30)

where M_{eff} is an effective mass of the vibrating dumbbell and f_{eff} the reduced effective spring constant as outlined above. For the simple dumbbell model

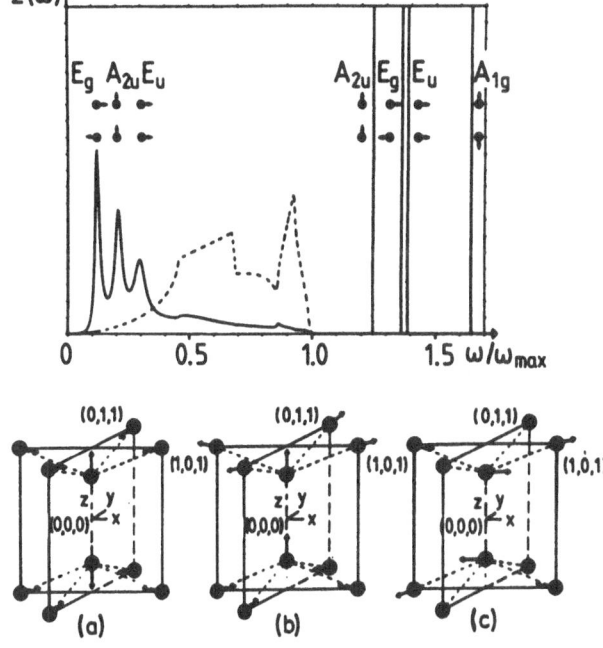

Fig. 2.10. Local frequency spectrum of a <100> dumbbell averaged over all directions for a Morse potential [3]. The vibration patterns are shown in more detail at the bottom of the figure

the corresponding vibration pattern would be similar to the arrows indicating the induced forces F_t. A more realistic example is shown in Fig. 2.10. It displays the results of a computer simulation for a [001] dumbbell in a fcc crystal with a Morse potential as interaction potential [3]. The local frequency spectrum exhibits three resonant vibrations at $\omega < 0.3\omega_{max}$. Here, ω_{max} is the maximum frequency of the spectrum of a normal lattice atom, which is also included in Fig. 2.10 for comparison as the dashed line. The displacement patterns of the respective vibrations are shown in the lower part of Fig. 2.10 in more detail. The mode with the lowest frequency, E_g, is related to the deformation introduced in the simple mechanical model above. The other two resonant modes are mainly related to the reduced spring constants acting between the dumbbell and their neighboring atoms indicating that they contribute additionally to the total polarizability. At frequencies above ω_{max} the spectrum exhibits so-called localized modes. One of the corresponding vibration patterns is indicated in Fig. 2.10 (A_{lg}, Fig. 2.10a). It is found that the dumbbell atoms and the neighboring atoms move in anti-phase in contrast to their in-phase motion during the resonant vibrations. Under the localized modes the springs interconnecting the atoms are stretched. Because of the anharmonicity of the potential $V(r)$, the spring constants $f = (\partial^2 V/\partial r^2)|_{r<R}$ are larger at distances $r < R_0$, than at the equilibrium distance R_0. The frequencies of the localized modes must be larger than ω_{max}.

The comparison of the symmetries of the resonant modes (Fig. 2.10) with the basic deformation modes (Fig. 2.6) allows qualitative relations between the respective diaelastic modulus changes to be derived. Because of the similarities between the vibrational patterns and the displacements it is apparent that for instance the librational mode E_g couples most strongly to the $\overset{[4]}{\underset{\sim}{\varepsilon}}$-type shear deformations. The $\overset{[2]}{\underset{\sim}{\varepsilon}}$ shear mode couples strongly to the localized mode A_{1g}, but only weakly to one of the resonant modes. From this one would expect that [11]

$$\overset{[4]}{\alpha}_d \gg \overset{[2]}{\alpha}_d$$

for the <100> dumbbell.

The sign of the polarizabilities is positive due to the fact that the effective restoring forces are reduced with respect to the normal force constants.

The preceding results indicate that the anisotropy of the diaelastic polarization is indicative of the symmetry of the local spring arrangement as modified by a point defect. Its measurement is therefore an experimental tool to obtain information on the configuration of a point defect. Thus it should for instance be possible to distinguish the <100> and the <110> oriented dumbbell.

The magnitude of the diaelastic effect is in principle related to the frequencies of the resonant modes. These frequencies and the respective weighting factors with which they contribute to the total effect depend sensitively on the interaction potential. Therefore it is difficult to deduce absolute values for the resonant frequencies from experimentally observed diaelastic modulus changes. However, as the resonant modes mostly modify the long wavelength part of the acoustic spectrum, one expects a change of slope of the phonon dispersion curves at long wavelengths, which can be observed for instance in neutron scattering experiments. Such results may then be directly compared to the magnitude and symmetry of the diaelastic modulus effects.

2.2 Elastic Interaction Between Point Defects

2.2.1 Paraelastic Interaction (Size Interaction)

A point defect (1) with dipole tensor $\underline{P}(1)$ at site $\boldsymbol{r}_1$ will interact with another defect (2) at a distance $\boldsymbol{R}$ via its strain $\underline{\varepsilon}_2$ according to (2.11) as

$$E^{\text{int}} = -(\underline{P}^{(1)}, \underline{\varepsilon}_2(\boldsymbol{R})) \sim \frac{1}{|\boldsymbol{R}|^3} \tag{2.31}$$

where $E^{\text{int}} = 0$ for isotropic defects in an isotropic material [1]. For all other cases E^{int} depends on the lattice direction of $\boldsymbol{R}$, but such that the average over the solid angle at fixed distances of the two defects is zero:

$$\langle E^{\text{int}} \rangle_{\vartheta\varphi} = 0 \,,$$

i.e. there are always directions of attractive and of repulsive interactions [1].

The preceding statements are strictly correct only for defects whose distance from the crystal surface is large compared to their mutual separations $|R|$. Otherwise, the displacement pattern $s(R)$ is modified by the so-called image forces giving rise to another interaction term [1]. This size interaction plays a dominant role in the formation of multiple interstitials. If interstitials are mobile, they may follow a path of attractive interaction, approach each other and ultimately be bound together in a multiple interstitial cluster.

2.2.2 Diaelastic Interaction (Inhomogeneity Interaction)

The expression for the interaction energy given in (2.31) is only complete if one takes into account the permanent and the induced dipole moments. The induced elastic dipole moment alone gives rise to an interaction term

$$E_{\text{ind}}^{\text{int}} = -(\underline{\alpha}_1 \underline{\varepsilon}_1, \, \underline{\varepsilon}_2(R)) \,. \tag{2.32}$$

The diaelastic interaction decreases with increasing distance $|R|$ typically more rapidly than does the paraelastic interaction. Therefore it is only important at small distances between the interacting species.

One example where the diaelastic interaction plays a leading role is the so-called "SIPA" mechanism, i.e. *stress-induced preferential absorption* of vacancies and interstitials at dislocations [23]. It occurs for instance in irradiated materials subjected to mechanical stresses. The diaelastic interaction is responsible for a coupling between the strain created by the dislocations, $\underline{\varepsilon}^{\text{dis}}$, and the external strains, $\underline{\varepsilon}^{\text{ext}}$:

$$E_{\text{ind}}^{\text{int}} = (\underline{\alpha}\,\underline{\varepsilon}^{\text{ext}}, \, \underline{\varepsilon}^{\text{dis}}) \,.$$

Dislocations with different orientations with respect to the external strain will possess different interaction energies with diffusing point defects. In this manner the anisotropic absorption rates of point defects at dislocations observed in irradiated materials may quantitatively be accounted for [23, 24, 160].

2.3 Dynamics and Geometry of Interstitial Jumps

2.3.1 Jumps in a Sinusoidal Potential

The dynamics of interstitial jumps may be simulated by a model where an effective mass, m, moves in a sinusoidal potential given by [161].

$$V(x) = \tfrac{1}{2}E_{\text{a}}[1 + \cos(\pi x/b)] \tag{2.33}$$

$$E_{\text{a}} = 8m\nu_0^2 b^2 \,. \tag{2.34}$$

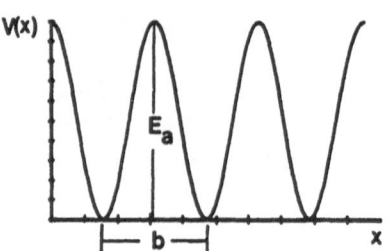

Fig. 2.11. Sinusoidal potential with activation energy, E_a and jump distance, b

Here E_a is the absolute height of the potential (Fig. 2.11). b is the period length and ν_0 the vibrational frequency given by the curvature at the potential minimum.

The classical transition rate of the particle from one minimum to the next is given by

$$\nu = \nu_0 \exp(-E_a/kT) . \tag{2.35}$$

If interstitial jumps are induced via the excitation of resonant modes, one may identify

$$\omega = \omega_R = 2\pi\nu_0$$

which is about one or two orders of magnitude smaller than for instance the corresponding preexponential factor observed for vacancy jumps during self-diffusion which is of the order of $10^{14}\,s^{-1}$.

A low vibrational frequency, ν_0, implies for a sinusoidal potential that the activation energy, E_a, as given by (2.34) is also low. In the same manner a small jump distance, b, also leads to a reduced activation energy.

2.3.2 Jump Modes of Self-Interstitials

Because of the dominance of the resonant vibrations in the local vibrational spectrum of the self-interstitials and the reasons given above, the resonant vibrations will govern the jump characteristics of the self-interstitials. In accordance with the displacement pattern of the <100> dumbbell, for instance, we may anticipate three independent jump modes as shown in Fig. 2.12: a pure rotation due to the librational (E_g) mode (Fig. 2.12a), and two translational modes, due to the A_{2u} (Fig. 2.12b) and the E_u mode (not shown here), respectively. One may omit the last one, however, from the further discussion because it would possess a particularly high saddle point energy [3].

In addition to the pure modes, one may also consider superpositions, e.g. of the E_g and A_{2u} modes. This combination leads to a simultaneous reorientation and migration of the self-interstitial as shown in Fig. 2.12c. From the simple argument given in Sect. 2.3.1 one would expect this jump to be the most favored one. This follows from an estimate of the jump distances. If one takes as a measure of the jump distance the net distances between

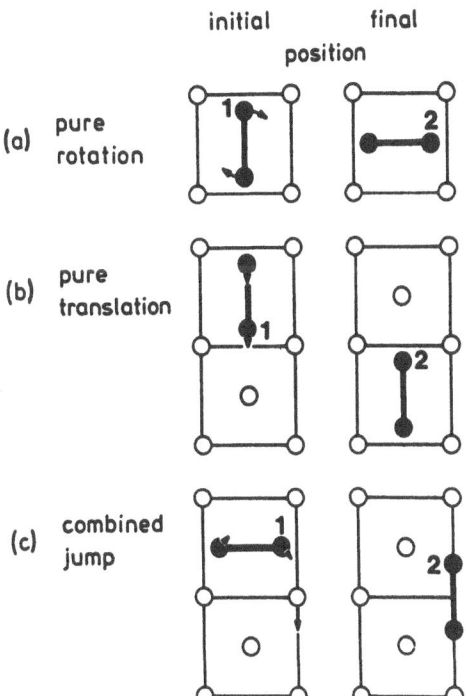

initial   final

position

(a) pure rotation

(b) pure translation

(c) combined jump

Fig. 2.12a–c. Three jump modes of the <100> self-interstitial atom in an fcc lattice: pure rotation, pure translation, and the combination of the first two

the respective atomic positions (1) and (2) before and after each of the three jumps of Fig. 2.12, one finds $b = \sqrt{2} \cdot (0.2a)$ for the combined jump, $b = 2 \cdot (0.2a)$ for the translational jump and $b = \sqrt{2} \cdot (0.3a)$ for the rotational jump, if the dumbbell distance is taken as $d = 0.6a$. Although this reasoning is not conclusive by itself, it may be regarded as a plausible argument for the result of the computer simulations, which lead to the same conclusion. This will be shown in Sects. 2.6, 7.

Another example of an interstitial jump where small jump distances and low frequency vibrational modes are involved is the so-called caging jump of the mixed dumbbell. As shown in Fig. 2.13, inside the octahedron where the solute atom is located there are six equivalent positions (see Fig. 2.19) which the solute atom may occupy. Jumps from one site to a neighboring one are again related to a combination of librational and translational vibrations of the dumbbell. In contrast to the capabilities of the pure solvent dumbbell and

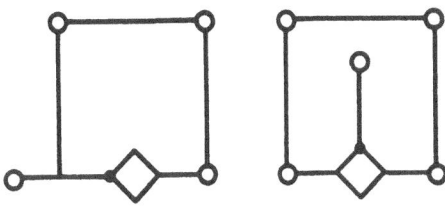

Fig. 2.13. Caging motion of the <100> mixed dumbbell. The jump of the solute atom to a neighboring site induces a reorientation of the axis of the mixed dumbbell

owing to the asymmetry of the mixed dumbbell structure, the jumps of the solute atom do not go beyond the limits imposed by the octahedral cell. In other words, the solute atom jumps around as if in a cage, to which the term caging jumps refers. The important difference between the mixed dumbbell and the pure solvent dumbbell jumps resides in the fact that the caging jumps give rise to a localized reorientation of the dumbbell axis, whereas in the second case the reorientation of the dumbbell axis is accompanied by a long-range migration of the interstitial configuration.

2.4 Time Dependence of the Paraelastic Polarizabilities: Relaxation Modes

Equation (2.20) demonstrates that self-interstitials can give rise to a para-elastic effect only if they possess an anisotropic double force tensor. A second necessary requirement for this effect to occur is that the self-interstitials are able to reorient from one orientation into the others in order to establish the altered equilibrium distribution (2.12). This is different from the diaelastic polarization where a defect motion is not required. The reorientation of the <100> dumbbell has already been exemplified in the previous Sect. 2.3.2 and we will use this example further to demonstrate the kinetics of self-interstitial reorientation. As described in Sect. 2.3.1 the transition rate from one orientation of the dumbbell into any of the other two is given by a jump rate (in the absence of external stress, Table 2.1)

$$\nu_{12} = \nu_{13} = \nu_{21} = \nu_{31} = \nu_{32} = \nu_{23} = \nu_0 \exp(-E^{\mathrm{R}}/kT) \ .$$

The subscripts indicate transitions of dipole axis 1 from orientation i into orientation j. ν_0 is the preexponential factor and E^{R} the activation energy for the reorientation. At the moment it is of no concern whether this reorientation occurs with or without simultaneous motion to another lattice site, as discussed above. We will simplify the present example by neglecting the third orientation and dealing only with two orientations denoted by 1 and 2. Figure 2.14 displays the energy contour between the two orientations for zero and non-zero deformation. If $\varepsilon \neq 0$, the energies of the two orientations split which results in a change of the net transition rates:

$$\nu_{12} = \nu e^{(\Delta W/2kT)} \tag{2.36a}$$

$$\nu_{21} = \nu e^{(-\Delta W/2kT)} \tag{2.36b}$$

$$\nu = \nu_0 e^{(-E^{\mathrm{R}}/kT)} \tag{2.37}$$

with $\Delta W = W^{(2)} - W^{(1)}$ as defined in (2.11).

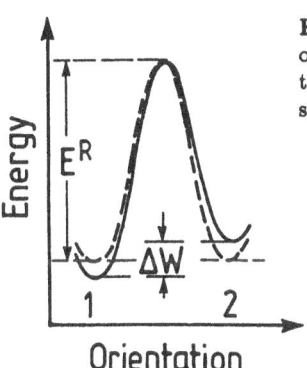

Fig. 2.14. Schematic energy contour between two defect orientations (1) and (2). E^R is the activation energy of the reorientational transition, and ΔW the stress-induced splitting of the elastic energy levels (2.11)

The concentration changes may be described by the rate equations

$$\dot{\varrho}_1 = \nu_{12}\varrho_2 - \nu_{21}\varrho_1 = -\dot{\varrho}_2 \, . \tag{2.38}$$

If one expands ν_{12} and ν_{21} to terms linear in $\Delta W/kT$ one obtains

$$\dot{\varrho}_1 = -2\nu(\varrho_1 - \bar{\varrho}_1) \tag{2.39}$$

where

$$\bar{\varrho}_1 = \varrho_0(1 + \Delta W/2kT)/2 \tag{2.40}$$

is the thermal equilibrium concentration of ϱ_1.

Equation (2.39) is the equation of a simple exponential relaxation process with a relaxation rate $\tau^{-1} = 2\nu$, where ν is the atomic jump rate involved in the underlying strainfree diffusion process. Since the polarization stresses $\Delta\overset{[\nu]}{\sigma}$, and the polarizabilities, $\overset{[\nu]}{\alpha}$, are proportional to $\varrho_1(t)$, they become explicitly time-dependent.

The time derivative of (2.4) and (2.38, 39) combined with the stress-strain relation

$$\sigma + \Delta\sigma = M\varepsilon$$

finally yield the equation of motion of a solid with paraelastic response:

$$\tau \cdot \dot{\sigma} + \sigma = M_R\varepsilon + \tau M_u \cdot \dot{\varepsilon} \, . \tag{2.41}$$

This equation is commonly known as the equation of motion of a standard anelastic solid [12,13]. It is a "time-dependent" Hooke's law. It is generally valid for any defect and any crystal symmetry. M_u stands for the elastic constants of the ideal "unrelaxed" crystal. M_R for the relaxed ones: $M_R = M_u + \Delta M$; $\Delta M = -\varrho_0\alpha$.

In detail the relaxation times, τ, are different for different relaxation modes, and they depend on the symmetries of the host lattice, of the dipole tensor and of the applied stresses [12]. In cubic crystals, they can be related again to the C and C' shear-type deformations, e.g. τ_C and $\tau_{C'}$, which can be expressed in terms of the elementary transition rates ν_{ij}. For the simple defect symmetries given in Table 2.1, the respective expressions are given in column 4 and 5 along with the corresponding polarizabilities, as taken from [12].

2.5 Quantum Effects

Several cases are known where interstitials are mobile at very low temperatures, typically around 1 K. This is for instance true of self-interstitials in Au, V, Nb, Ta, Cd, Mg and in some alloys like Al–Fe and Al–Zn [2]. Evidence exists that the motion occurs via tunnelling processes rather than via classical "hopping" motions described by an Arrhenius type behavior.

Such quantum effects may be described in terms of the simple model employing a one-dimensional sinusoidal potential as before. The system is described by two wave functions, which are symmetric and antisymmetric with respect to two adjacent potential minima. The two states possess energies which differ by an amount ΔE, the so-called tunnel splitting. The tunnel frequency between the two sites is given by

$$\nu_T = \Delta E / h \tag{2.42}$$

where h is Planck's constant. The sinusoidal potential defined in (2.33) yields [161] for $\eta \ll 1$

$$\Delta E = \frac{2}{\pi^{3/2}} h\omega \sqrt{\eta} \exp[-(8/\pi^2)\eta] \tag{2.43}$$

where

$$\eta = \frac{m\omega^2 b^2}{\hbar\omega} , \quad \hbar = h/2\pi . \tag{2.44}$$

Figure 2.15 depicts the situation.

Equation (2.43) shows that ν_T depends very sensitively on the jump distance, b, as $\sim\exp(-m \cdot b^2 \cdot \text{const})$, and less sensitively on the mass, m.

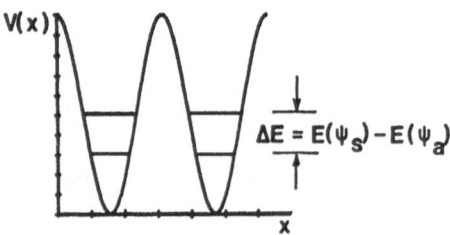

Fig. 2.15. Simple model for tunnelling transitions between two orientations (lattice positions) via tunnelling motions. ψ_s and ψ_a are the symmetrical and antisymmetrical wave functions, respectively

As an example, *Schober* and *Stoneham* [161] calculate a tunnel frequency of $\nu_T = 10^{-17}\,s^{-1}$ for a <100> dumbbell in Al (case 1a of Table IV in [161]), but $\nu_T = 10^4 - 10^8\,s^{-1}$ for a mixed dumbbell in Al–Fe. In other words, tunnelling motions of self-interstitials are possible but difficult to forecast theoretically for a specific case owing to the sensitivity of the tunnelling rate to details of the atomic jump.

2.6 Computer Simulations of Single and Multiple Self-Interstitials

The aim of the preceding chapters was to emphasize the principle concept by which para- and diaelastic phenomena may be described. Several examples were addressed, however, which were obtained from computer simulations, describing the defects and their mechanical response in more detail. Such results are very valuable for comparison with experimental results. For this reason a short summary of this subject will be given.

Computer simulations are done in the following manner [2]. The defect to be studied is embedded in a crystallite of some thousand atoms. The crystallite itself is placed into a larger crystal. The atoms of the crystallite are allowed to relax under the influence of their mutual interactions and their interaction with the atoms of the surrounding larger crystal, where the atoms of the latter are fixed to their ideal crystal sites. The stable configurations of the defects are found from the minimization of the potential energy of the crystallite. Activation energies are obtained from the potential energies of the saddle point configurations, which are obtained by the same procedure including additional geometrical constraints.

The results of such calculations, e.g. as given by *Dederichs* et al. [3], are summarized in Table 2.2. The potentials used in these examples were a modified Morse potential for Cu and a special bcc potential for Fe ("Johnson's potential for α-Fe"). Table 2.2 gives values for the formation energies, E^F, relaxation volumes, trace of dipole tensors, $Tr\underline{P}$, normalized eigenvalues of the dipole tensors, P_α, and the principle axes e_α. The configurations listed are the [001] dumbbell and its <100> saddle point (fcc), the [110] dumbbell, the [111] dumbbell, and the "diffusion saddle", (bcc). Computer simulation may also be employed to deduce diaelastic polarizabilities. This can be done deforming the crystallite containing the defect, and comparing its deformation energy with the energy of a deformed crystallite without defect. In this manner, the data of Table 2.3 were obtained. Here, the respective diaelastic modulus changes are given in terms of an averaged shear modulus $G = 0.6\,C + 0.4\,C'$. The data of Table 2.2 and 2.3 quantify the more qualitative discussion of the preceding chapters. The general trends may be summarized as follows:

The formation energies of self-interstitials are considerably higher than those of vacancies. The relaxation volumes of self-interstitials are larger and exceed typically one atomic volume. The dipole tensor of the [100] dumbbell is

Table 2.2. Summary of computersimulation data: formation energies (E^F), volume change V^{rel}, trace of dipole tensors, Tr $\underline{P}$, eigenvalues of dipole tensors P and main axes e. Modified Morse potential (fcc), Johnson's Fe-potential (bcc), after [3]

Configurations	E^F [eV]	V^{rel}	Tr (P/eV)	P_1	P_2	P_3	e_1	e_2	e_3	crystal
001 dumbbell	3.42	1.5	44.4	1.02	1.02	0.96	(1,0,0)	(0,1,0)	(0,0,1)	fcc
100 migration Saddle point	3.55	1.53	45.3	1.17	0.78	1.05	(1,0,1)	(1,0,$\bar{1}$)	(0,1,0)	fcc
110 dumbbell	4.6	2.34	62.4	1.29	0.61	1.10	(1,1,0)	(1,$\bar{1}$,0)	(0,0,1)	bcc
111 dumbbell*	4.88	2.20	61.7	1.94	0.53	0.53	(1,1,1)	(1,$\bar{1}$,0)	(1,1,$\bar{2}$)	bcc
"Diffusion saddle" Point*	4.81	2.25	63.1	1.50	0.89	0.61	(1,1.05,1)	(1,0,$\bar{1}$)	(1,−1.9,1)	bcc
Vacancy	1.29	−0.02		−0.22	−0.22	−0.22	(1,0,0)	(0,1,0)	(0,0,1)	fcc
Vacancy, Saddle point	2.22	0.16		−1.57	0.49	5.75	(1,1,0)	(1,$\bar{1}$,0)	(0,0,1)	fcc

* The <111> dumbbell may be regarded as a saddle point configuration for a second jump mode, discussed in Sect. 4.2.2.

Table 2.3. Change of elastic constants due to Frenkel defects. G is the averaged, isotropic shear modulus of Voigt $G = (3/5)C_{44} + (2/5)(C_{11} - C_{12})/2$. The calculations refer to three different potentials supposed to resemble Cu. From [3]

			$\frac{\Delta C_{44}}{cG}$	$\frac{\Delta(C_{11}-C_{12})/2}{cG}$	$\frac{\Delta(C_{11}+2C_{12})}{c(C_{11}+2C_{12})}$
Theory	<100> dumbbell	BM	−34.6	1.2	−1.2
		MO	−50.0	−4.8	−8.9
		MM	−21.8	−2.3	−8.7
	vacancy	BM	−5.7	−5.7	−0
		MO	−4.2	−2.0	−2.7
		MM	−3.3	−2.0	−2.4
	Frenkel pair	BM	−40.3	−4.5	−1.2
		MO	−54.2	−6.8	−11.5
		MM	−25.1	−4.3	−11.4

almost cubic, the anisotropic deviation amounts to a few percent. In contrast to this, the bcc configuration is strongly anisotropic. The activation energies for the interstitial jumps are small compared to the energies for vacancy migration, for instance 0.13 eV compared to 0.93 eV in the fcc case.

Self-interstitial clustering has been investigated by *Ingle* et al. [4]. The configurations and their respective formation energies were determined from computer simulation. Figure 2.16 displays some examples from this study. The stable di-interstitial (a) consists of a pair of dumbbells centered on nearest neighbor sites. The parallel dumbbell axes are tilted slightly off the exact <100> direction. This configuration was found to possess an energy 0.1 eV smaller than a pair of exactly <100>-oriented dumbbells. The symmetry is <110> orthorhombic. The tri-interstitial Fig. 2.16b, consists of 3 mutually orthogonal <100> dumbbells on nearest neighbor sites, the tetra-interstitial Fig. 2.16c, has a similar structure except for an additional atom on the octahedral center. Both configurations, the tri- and tetra-interstitial, possess <111> trigonal symmetry. Larger three-dimensional clusters follow from the tetra-interstitial by successively replacing the face-centered atoms by <100> dumbbells until a structure with seven interstitials is reached as shown in Fig. 2.16e, with cubic symmetry. Even larger three-dimensional clusters can be derived from the seven-interstitial by further adding octahedral or dumbbell interstitials.

Alternatively, also planar or two-dimensional arrangements of self-interstitials are possible. As an example a planar seven-interstitial is shown in Fig. 2.16f.

The formation energy E_f/N per interstitial in a cluster of N interstitials as a function of cluster size N is shown in Fig. 2.17 for two- and three-dimensional arrangements, marked 2D and 3D, respectively. A transition point occurs at about $N = 10$. For $N < 10$, 3D clusters seem to be preferred and 2D clusters for $N > 10$.

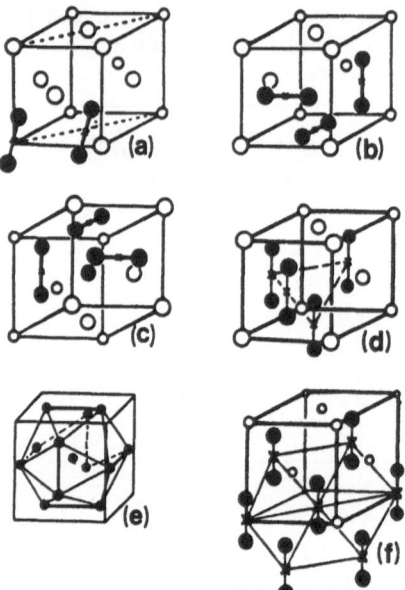

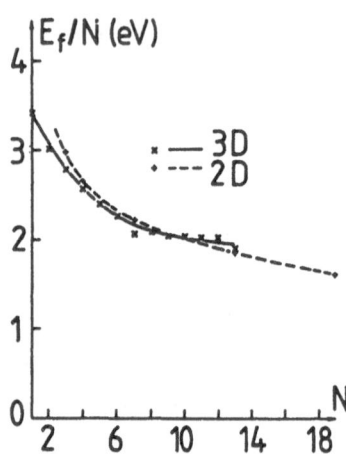

Fig. 2.16. Three- and two-dimensional interstitial clusters in an fcc lattice, after [4]

Fig. 2.17. Formation energy per interstitial in a cluster containing N interstitials as a function of cluster size N (Morse potential, [4]) 2D and 3D refer to two- and three-dimensional entities

The only well defined cluster identified in a bcc lattice is the di-interstitial atom [5]. It consist of two parallel <100> dumbbell atoms with a binding energy of about 1 eV. The nucleation and growth of interstitial loops starting from the <110> split configuration has been studied by *Bullough* and *Perrin* [6]. It has not been determined whether other small three-dimensional clusters may exist.

2.7 Computer Simulations of the Interaction of Self-Interstitials with Solute Atoms

2.7.1 Structure of SI Complexes

The reaction between self-interstitials and solute atoms will lead to the formation of stable complexes if there is a reduction of the configurational energy associated with it. It will turn out that this is the case for a large variety of different atomic structures. Therefore the general term SI complexes will be used (short for *solute – self-interstitial* complex) as long as specific configurations are not meant.

Dederichs et al. [3] have proposed a simple model for estimating the energy gained by replacing a host atom of the dumbbell by a solute atom. In

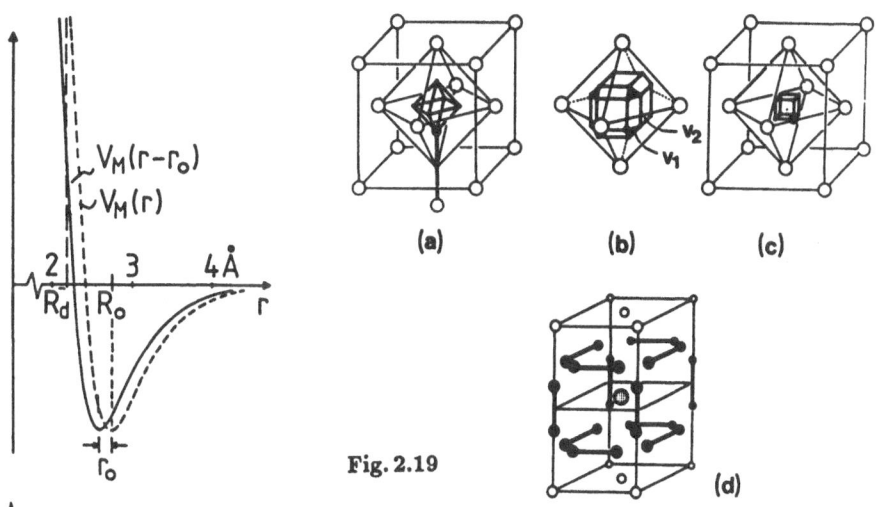

Fig. 2.18. A regular Morse potential representing host-host atom interaction ($V_M(r)$) and the same Morse potential shifted by an amount r_0 used to simulate solute-host atom interaction ($V_M(r - r_0)$). R_0 is the regular lattice separation of the host atoms, R_d is the typically smaller dumbbell atom separation

Fig. 2.19. Atomistic models of SI complexes derived from the simulations of the solute-host atom interaction by a shifted Morse potential (see Fig. 2.18): (a) mixed dumbbell, (b) split cubic cage, (c) simple cubic cage, (d) 12 equivalent nearest neighbor positions of a regular dumbbell next to an oversized solute atom on a substitutional site

this model the interaction of an undersized solute atom with its neighbors is represented by a pair potential which was obtained by shifting the repulsion part of the interaction potential by an amount r_0 to smaller r values with respect to the original Morse potential representing the host atom-host atom interaction (Fig. 2.18). With this simple model it was found that differently structured SI complexes may occur for differently sized solute atoms: for $0 \leq -r_0/R_0 \leq 0.06$, where R_0 denotes the lattice equilibrium distance, the so-called <100> mixed dumbbell is the most stable configuration (Fig. 2.19a). For $-r_0/R_0 > 0.07$, the solute atom prefers other sites near the octahedral position, for instance one of the cube corners shown in Fig. 2.19c. For $-r_0/R_0 > 0.09$, the solute atom occupies the octahedral center. Oversized solute atoms are not expected to be incorporated as a dumbbell member, because they would increase the lattice expansion. Instead, the oversized solute atom occupies a nearly substitutional site next to a regular host atom-host atom dumbbell on one of the 12 equivalent sites shown in Fig. 2.19d (and Fig. 2.3 earlier).

A calculation taking the electronic structure into account explicitly has been given by *Lam* et al. [67]. Using pair potentials derived from pseudo-potential theory, they found, e.g., the <100> mixed dumbbell (Fig. 2.19a) to be the most stable complex in A̱l–Zn, and the cubic cage Fig. 2.19c in A̱l–Li

[137]. Other Al alloys investigated by Lam et al. include A̲l–Ag, Au, Be, Ca, K, Mg.

The simulation of the host-solute atom-interaction by a shifted potential is a severe simplification. It assumes that the main part of the interaction energy is due to the strong repulsive part of the potential at distances smaller than the nearest neighbor distance. This concept works well in the case of self-interstitials but it does not in the case of substitutional positions, where the long-range oscillations of the interaction potential are as important as the short-range repulsive part [65]. For this reason the shift parameter, r_0, cannot be derived unambiguously from lattice parameter measurements on substitutional alloys. The structure of the SI complexes is very sensitive to the form of potential used to simulate the self-interstitial – solute atom interaction [66]. An example is shown in Fig. 2.19b, where the solute atom occupies one of the 24 equivalent sites around the octahedral center. This configuration was obtained by empirically introducing a slight additional oscillation into the interaction potential. The symmetries and widths of the resulting splittings of the octahedral center were found to depend very sensitively on the position as well as on the height of the oscillation.

In conclusion, a reliable simple parameter which allows a prediction of the SI complex configuration for a given alloy has not been found. However, as a rule of thumb solute atoms which are undersized with respect to the host atoms are transferred into interstitial sites whereas an oversized solute atom will trap a self-interstitial on a nearest neighbor position preserving the pure host atom-host atom dumbbell.

2.7.2 Binding Energies

The octahedral site and the "split octahedral" configurations of Fig. 2.19 are not the only stable positions on which undersized solute atoms can be accommodated. Figure 2.20a shows for instance nearest and several next nearest neighbor sites of a solute atom to the dumbbell with the respective binding energies. These binding energies are given in units of the binding energy of the mixed dumbbell. Positive numbers indicate attractive interaction, i. e. stable positions, and negative numbers unstable positions. (The binding energies scale with the shift parameter, r_0, is shown in Fig. 2.20b.) For this reason, the sign of the various binding energies of Fig. 2.20a changes when the undersized solute atom is replaced by an oversized solute atom, i. e. stable positions become unstable and vice versa. The nearest neighbor position at $\frac{1}{2}(110)$ to the dumbbell at (000) is considered to be the most stable form of a SI complex involving an oversized solute atom (cf. Fig. 2.19d). Multiple trapping, i. e. binding of more than one self-interstitial at one solute atom, can also occur. The calculations by *Schober* et al. [65] show that these larger trapped complexes closely resemble the configurations of multiple interstitials in pure metals [68–70] (Fig. 2.16). They consist of either parallel or mutually perpendicular dumbbells, which are grouped around a mixed dumbbell or

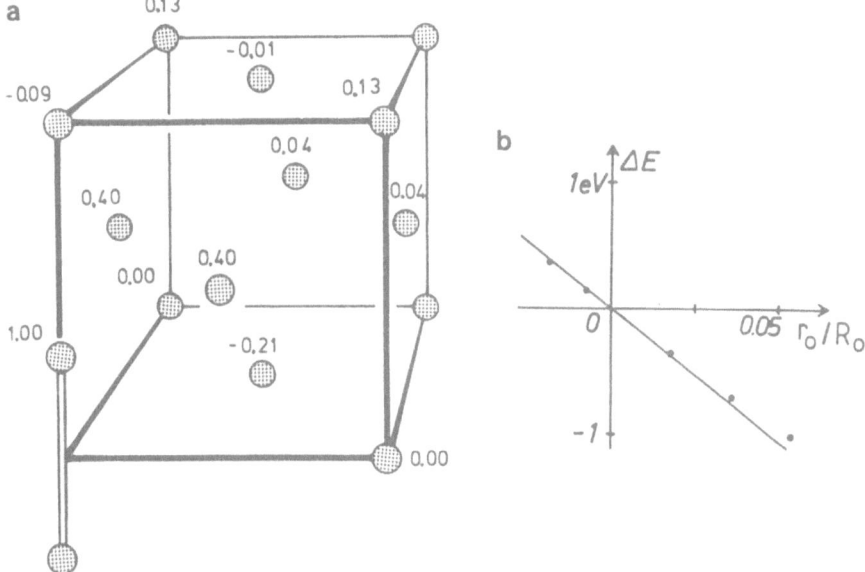

Fig. 2.20. (a) Binding energies, $\Delta E/\Delta E_{md}$, for the various SI complexes, in units of the binding energy of the mixed dumbbell, ΔE_{md}. An undersized solute atom ($r_0 < 0$) is placed at the respective lattice size. Positive numbers correspond to attractive configurations for $r_0 < 0$, and vice versa. The numbers are for the Morse potential of Fig. 2.18. (b) The binding energy, ΔE_{md}, of the mixed dumbbell as a function of the potential shift, r_0

an oversized substitutional solute atom. The binding energy per additional self-interstitial is of the order of 1 eV and increases slightly with the size of the complex. The kinetics of the build-up of single and multiple SI complexes under irradiation has been simulated using rate equations by *Bewerunge* [71] and *Becker* et al. [72].

Lam et al. [67, 137, 138] have also obtained binding energies of various SI complexes based upon their electronic calculations. Since these are given for specific alloys, they will be discussed along with the experimental results.

2.7.3 Mobilities

Similarly to the self-interstitials, the SI complexes may also perform thermally activated jumps to neighboring positions. In the following we will discuss jumps of the mixed dumbbell as representatives of each of the configurations of Fig. 2.19a–c, and the jumps of the bound dumbbell around a solute atom. Four different jump processes of the mixed dumbbell may occur. As shown in Fig. 2.21a–d these are: caging-, rotation-, dissociation and "looping" jumps. As mentioned earlier, caging jumps (Fig. 2.13) occur when the solute atom jumps to one of the other equivalent positions within the cage. Each of these jumps is connected with a change of the host atom partner and thus with a reorientation of the mixed dumbbell axis. The diffusion is localized, the

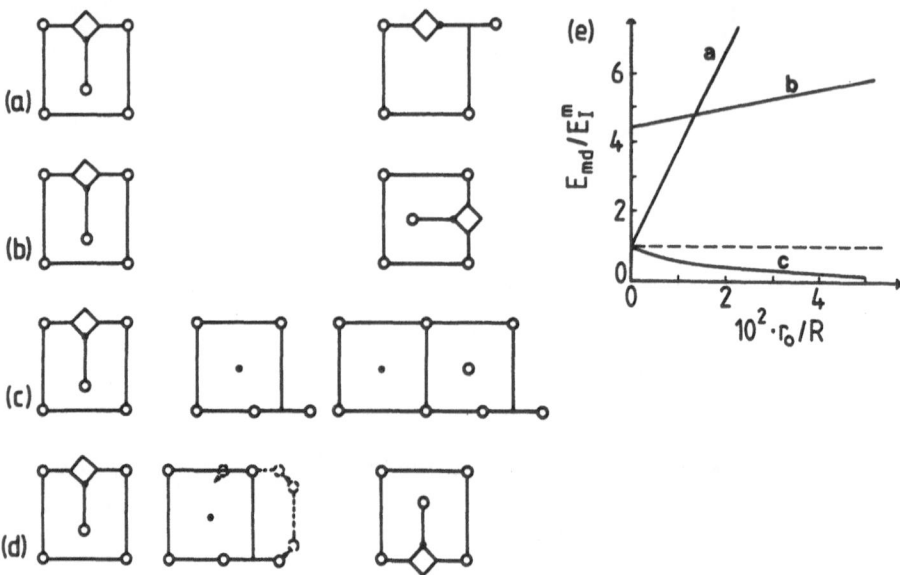

Fig. 2.21a–d. Jump modes of the mixed dumbbell: (a) caging, (b) rotation, (c) dissociation, (d) looping. (e) Activation energies of the various jump modes of the mixed dumbbell as a function of the shift parameter, r_0: (a) dissociation, (b) rotation and (c) caging. Figure from [3]

solute atom moves as if in a cage. The distance from one position to the next is small compared to a nearest neighbor distance of the lattice. According to (2.43) and (2.44) we therefore expect a low activation energy for this process. The rotation jumps occur as an in situ rotation of the whole mixed dumbbell (Fig. 2.21b). Dissociation jumps occur when the host atom partner of the mixed dumbbell jumps away from the solute atom and forms a pure host atom dumbbell whereby the solute atom returns to a substitutional position (Fig. 2.21c). The activation energy of this process is the sum of the binding energy of the SI complex, ΔE, and the migration energy, E_I^m, of the self-interstitial. The activation energies of the various jump processes depend on the shift parameter r_0 as shown in Fig. 2.21e. The caging jumps of the mixed dumbbell have the lowest activation energy, which decays with increasing size factor. In contrast to this , the dissociation (Fig. 2.20b) and rotation energies (Fig. 2.20a) increase with increasing size factor. For small size factors, e.g. $-r_0/R_0 < 0.015$ in Fig. 2.20, the dissociation jump is easier to excite than the rotation. Therefore solute atoms in this range can, if at all, only be transported via a looping mechanism. Above this point, transport via the caging and rotation combination becomes favorable, since many of these jumps occur before a dissociation jump destroys the mixed dumbbell.

A reassociation of the self-interstitial with "its" solute atom may occur owing to some still acting interaction, as for instance shown in Fig. 2.21d, where the dumbbell returns as if in a loop.

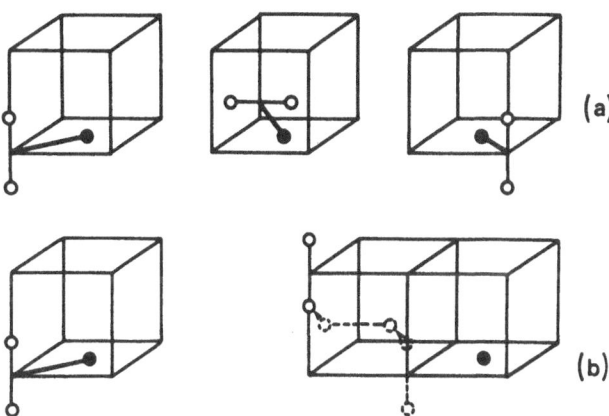

Fig. 2.22. (a) Reorientation of the nearest neighbor complex by the combined jump of the regular dumbbell. (b) Dissociation of the nearest neighbor complex

If a regular dumbbell is bound to a solute atom on a neighboring position as for instance in Fig. 2.19d, the dumbbell has access to several equivalent positions grouped around the solute atom, e. g. the 12 sites shown in Fig. 2,19d. Jumps between these sites are possible by the normal diffusional jump of the dumbbell as shown in Fig. 2.22a. If the binding energy can be overcome, a dissociation of the complex may occur (Fig. 2.22b).

Preferential transport of solute atoms may, within the preceding models of Fig. 2.21, occur by two mechanisms. The first is a combination of the caging and rotation motions. The rotation carries the solute atom into a neighboring cage. From there it may reach a new site in this cage by the caging motion. From these positions another cage may be reached by a second rotation and so forth. The second mechanism of long-range solute atom transport is the looping mechanism, by which the solute atoms are shifted into other cages. SI complexes in other configurations, e. g. as in Fig. 2.19b and c, may in principle exhibit jumps analogous to the ones discussed before [137, 138].

In principle, "cages" are also possible in the bcc structure, but there is not much information from theoretical studies. One may "construct" cages for the bcc structure following the procedures applied in the fcc case, namely by splitting the octahedral center into highly symmetric directions. As an example, the analogon to the fcc mixed dumbbell is shown in Fig. 2.23 [158]. From the six sites in the octahedral arrangement, the four arranged in the square are equivalent, but different from the two remaining sites opposing the square. Therefore the "cage" established for the <110> mixed dumbbells is a plane square arrangement of only four equivalent sites. Jumps among the sites induce a 90° degree rotation of the mixed <110> dumbbell. There are two kinds of rotation jumps, as shown in Fig. 2.23c and d. Under the assumption that the activation energy scales with the jump distance squared, the latter jump would be preferred to the previous one. Also, the jump of Fig. 2.23d corresponds to the normal octahedral interstitial jumps, if the squares degen-

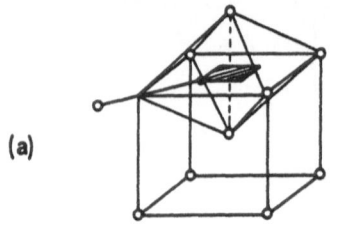

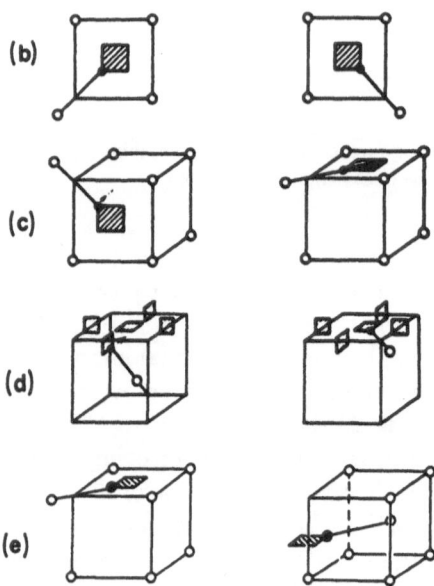

Fig. 2.23. Schematic outline of a possible cage configuration of a mixed <110> dumbbell in a bcc lattice (a), and jump modes (b) to (e) analogous to jump modes proposed for a mixed dumbbell in the fcc lattice. (b) displays the caging motion, (c) and (d) are two possible rotational jumps, and (e) a "shift" jump of the solute atom. The dissociation jumps have been omitted here

erate into a point. As for the fcc case, a combination of caging and rotation may give rise to preferential solute atom transport. Another mechanism for solute atom transport shown in Fig. 2.23 has been pointed out by *Lucasson* et al. [136]. The transport follows from a sequence of the shift-jumps, under the condition that a preferential shift to the neighboring position of the solute atoms occurs.

A dissociation of the bcc mixed dumbbell should also be possible.

In several alloys as for instance P̲b̲–Au a transport of the Au atoms is observed which is by several orders of magnitude faster than the self-diffusion in Pb. It is thought that also in these cases an interstitial mechanism is responsible for the mass transport. A fraction of the normally substitutional solute atom is present as highly mobile SI complexes in thermal equilibrium and these complexes dominate the impurity diffusion. It has not been established, however, to what extent the configurations of these SI complexes in the "fast diffusion" couples are related to the configurations observed under irradiation.

2.7.4 Trapping Radii

The probability of trapping a migrating self-interstitial at a solute atom is determined by the long-range part of their mutual interaction potential. *Schroeder* et al. [74] have given a method to express the reaction probability in terms of this interaction potential. As an example, Fig. 2.24 shows schematically an interaction potential leading to a drift diffusion of the self-interstitial towards and a subsequent trapping at the solute atom located at the origin. The trapping rate is expressed in terms of a trapping radius r_t as

$$K = 4\pi D_0 C_t r_t$$

where D_0 is the diffusion constant of the self-interstitials and C_t the concentration of solute atoms acting as trapping centers. Whenever a self-interstitial atom approaches the solute atom within a distance r_t, the self-interstitial atom starts a drift diffusion towards the solute atom, which inevitably leads to trapping. As seen in Fig. 2.24 "this point of no return" is reached, when the self-interstitial atom coming from infinity has gained an amount of potential energy in a saddle point configuration which equals the thermal energy kT.

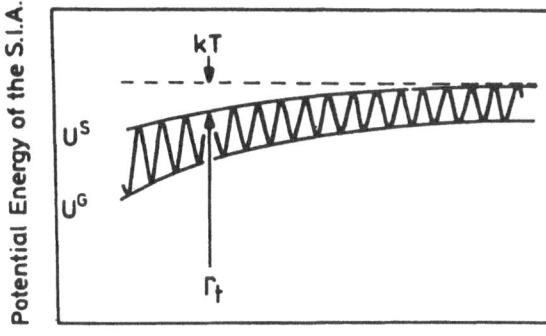

Fig. 2.24. Interaction potential of a self-interstitial with a solute atom as a sink. The trapping properties of the solute atom are characterized by the trapping radius r_t. It is defined by the condition that the potential energy gained by the self-interstitial in a potential saddle equals the thermal energy, kT. U^S and U^G refer to saddle and ground state, respectively

3. Experimental Techniques

Mechanical polarization experiments are designed to determine the para- and diaelastic response of a material containing the defects of interest. Accordingly the main quantities of interest are:

1) The magnitudes of the paraelastic polarizabilities or modulus changes, $\overset{[\nu]}{\alpha}$ or $\Delta\overset{[\nu]}{C}$;

2) the relaxation times, $\overset{[\nu]}{\tau}$;

3) the magnitudes of the diaelastic polarizabilities or modulus changes, $\overset{[\nu]}{\alpha}_d$ or $\Delta\overset{[\nu]}{C}_d$.

Experimental techniques capable of measuring these quantities will be discussed in the following. Two classes of experiments exist. The first class comprises experiments, in which the quasi-static response of a crystal is measured directly, i.e. strain or stress relaxation measurements. The second class comprises experiments, in which the dissipation of elastic energy due to the stress-induced defect reorientations is observed. With both techniques it is possible to fully characterize the para- and diaelastic properties.

3.1   Creep and Elastic Aftereffect

Following the nomenclature of *Nowick* and *Berry* [12] creep or strain relaxation is the time-dependent strain, $\varepsilon(t)$, exhibited by the tested material after an instantaneous application of a loading stress σ_0, cf. Fig. 3.1. $\varepsilon(t)/\sigma_0$ is called the creep function, $J(t)$. It has the dimension of an elastic compliance.

At $t = 0$, $J(0) = M^{-1}(0) = J_U = M_U^{-1}$, where J_U and M_U are called the unrelaxed compliance and modulus, respectively. M represents one of the elastic moduli $\overset{[\nu]}{C}$ introduced in Chap. 2. If the material exhibits diaelastic effects, the diaelastic modulus change, ΔM_d, is contained in M_U. This means that the diaelastic effect may be obtained from a comparison of samples with and without defects, respectively:

$$\Delta M_d = M_U \text{ (with defects)} - M_U \text{ (without defects)} .$$

Curve (a) of Fig. 3.1, for which $\varepsilon(t) = \varepsilon(0) = \text{const}$, represents ideal elastic

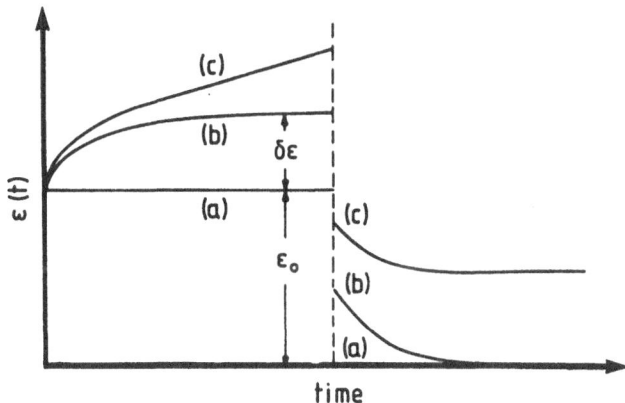

Fig. 3.1. Schematic representation of the creep and elastic aftereffects curve $\varepsilon(t)/\sigma_0$ for different classes of materials: (a) ideal elastic (b) paraelastic, (c) general visco-elastic case

behavior, i. e. zero strain relaxation. On the other hand, curve (c) represents a more general case where $\varepsilon(t)$ follows after a transient period, a linear trend. Such behavior is characteristic of a linear visco-elastic solid.

The case of purely paraelastic or anelastic behavior is represented by curve (b). It is distinguished from all other cases by the fact that in accordance with the definition in Sect. 2.1.3 an equilibrium value $\varepsilon(t \to \infty) = J_R \cdot \sigma_0 = \sigma_0/M_R$ exists towards which $\varepsilon(t)$ converges with increasing time

$$\Delta\varepsilon = \varepsilon(t \to \infty) - \varepsilon(t = 0) = \sigma_0 \left(\frac{1}{M_R} - \frac{1}{M_U} \right) \tag{3.1}$$

$$\Delta\varepsilon = -\varepsilon_0 \frac{\Delta M}{M_R}, \quad \text{with} \quad \varepsilon_0 = \varepsilon(t = 0), \quad \Delta M = M_R - M_U . \tag{3.2}$$

The quantity $\Delta_M = (M_U - M_R)/M_R$ is termed relaxation strength. The time evolution of the paraelastic polarization strain follows from (2.41) as

$$\delta\varepsilon(t)/\varepsilon_0 = \Delta_M \cdot [1 - \exp(-t/\tau_\sigma)] \quad \text{(loading)} \tag{3.3}$$

where $\delta\varepsilon(t) = \varepsilon(t) - \varepsilon_0$.

τ_σ is the relaxation time at constant stress ($\dot\sigma = 0$). If the loading stress σ_0 is instantaneously switched off at a time t_0, the total strain jumps back from $\varepsilon(t_0)$ to $\delta\varepsilon(t_0)$. If the loading time was sufficiently large, i. e. $t_0 \gg \tau_\sigma$ we have at the end of the loading period $\delta\varepsilon(t_0) \simeq \Delta\varepsilon$. For times $t > t_0$ the strain tends exponentially back to zero.

$$\delta\varepsilon(t) = \delta\varepsilon(t_0) \exp[-(t - t_0)/\tau_0] \quad \text{(unloading)} . \tag{3.4}$$

This part of the experiment is termed elastic aftereffect or creep recovery.

In an analogous manner experiments may be performed in which by adjusting the external stress, $\sigma(t)$, the strain, ε, is kept constant. These are called stress relaxation experiments. If relaxation processes obey an Arrhenius type behavior, a useful manner of representing relaxation data is by constructing so-called isochronal aftereffect curves. These are defined by

$$I(T) = [\delta\varepsilon(t_1, T) - \delta\varepsilon(t_2, T)]/\varepsilon_0 , \quad t_2 > t_1 > t_0 .$$

In such $I(T)$ plots relaxation processes reveal themselves as peaked functions, whose heights are proportional to the respective relaxation strengths, and whose positions are at a temperature at which the relaxation times are

$$\tau = (t_2 - t_1)/\ln t_2/t_1 .$$

Examples of such plots will be given in Chap. 4.

3.2 Elastic Energy Dissipation

If a solid with paraelastic properties is subjected to a vibrational strain, dissipation of elastic energy may occur due to the vibration-induced reorientation of the elastic dipoles in the material. As a result, characteristic damping phenomena are observed experimentally, for instance Internal Friction or Ultrasonic Attenuation as discussed below.

In order to elucidate the mechanisms by which elastic energy dissipation can occur, we solve (2.39) for an alternating strain:

$$\tau\dot{\varrho}_1 = -(\varrho_1 - \bar{\varrho}_1) ; \quad \bar{\varrho}_1 = \frac{\varrho_0}{2}\left(1 + \frac{\Delta P \cdot \varepsilon}{2kT}\right) ; \quad \varepsilon = \varepsilon_0 \sin\omega t . \tag{3.5}$$

By expressing the densities as deviations from their zero strain values

$$u_1 = \varrho_1 - \frac{\varrho_0}{2} \tag{3.6}$$

we obtain the first order inhomogeneous differential equation

$$\tau\dot{u}_1 + u_1 = \varrho_0 \frac{\Delta P \varepsilon_0}{4kT} \cdot \sin\omega t \tag{3.7}$$

with the solution [13]

$$u_1(t) = a(\sin\omega t - \omega\tau \cos\omega t)$$
$$a = \varrho_0 \frac{\Delta P \varepsilon_0}{4kT} \cdot \frac{1}{1 + \omega^2\tau^2} . \tag{3.8}$$

Equation (3.8) shows that the site population u_1 and correspondingly u_2 decompose into a fraction which moves in-phase with the vibrational strain and a second fraction which moves 90° out-of-phase. The ratio of the two fractions is given by $\omega\tau$, and they are shown in Figs. 3.2 and 3.3, respectively.

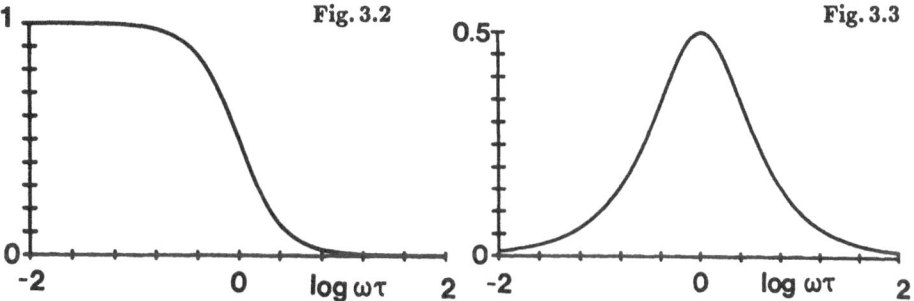

Fig. 3.2. Normalized fraction of defects in orientation 1 of the two-orientation model (Fig. 2.7) which moves in-phase with the vibrational strain $\varepsilon(t)$. The normalized modulus curves $(\omega'^2 - \omega^2)/\omega^2$ follow the same trend (3.15)

Fig. 3.3. Normalized fraction of defects in orientation 1 of the two-orientation model (Fig. 2.7) which moves $90°$ out-of-phase with respect to the vibrational strain. This curve describes at the same time the slope of the internal friction in units of the relaxation strength, i.e. Q^{-1}/Δ (3.13)

The energy dissipated in one vibrational cycle is given by

$$\Delta U = \int_0^{2\pi/\omega} \Delta W(\varepsilon)\dot{u}_1(t)\, dt$$

$$= \int_0^{2\pi/\omega} \Delta P \varepsilon_0 \sin\omega t \cdot a\omega(\cos\omega t + \omega\tau \sin\omega t)\, dt . \tag{3.9}$$

The first term in the bracket cancels out, i.e. the in-phase fraction does not contribute to the dissipation. The energy dissipation is caused entirely by the $90°$-phase component and is given by:

$$\Delta U = \frac{\pi(\Delta P \cdot \varepsilon_0)^2}{4kT} \cdot \varrho_0 \cdot \frac{\omega\tau}{1 + (\omega\tau)^2} . \tag{3.10}$$

This kind of response is called Debye peak. As shown in Fig. 3.3, ΔU is a peaked function at $\omega\tau = 1$ and tends to zero for $\omega\tau \gg 1$ and $\omega\tau \ll 1$. The reason for this behavior is clear: at low frequencies, i.e. $\omega \ll \tau^{-1}$, the defects can easily follow the vibration, so that the out-of-phase fraction tends to zero and therefore also the dissipated energy. For high frequencies, i.e. $\omega \gg \tau^{-1}$, where the reorientation rate is low compared to the vibrational frequency of the sample, the defects cannot follow the vibration at all. Therefore the dissipated energy tends to zero again. At $\omega\tau = 1$, the out-of-phase fraction and therefore the dissipated energy attain their maximum values.

It is convenient to express the dissipated energy in units of the maximum stored elastic energy of the vibration which is

39

$$U = \tfrac{1}{2}\sigma_0\varepsilon_0 = \tfrac{1}{2}M\varepsilon_0^2 . \tag{3.11}$$

The quantity

$$Q^{-1} = \frac{1}{2\pi} \cdot \frac{\Delta U}{U} = \frac{\varrho_0}{M} \cdot \frac{(\Delta P)^2}{4kT} \cdot \frac{\omega\tau}{1+(\omega\tau)^2} \tag{3.12}$$

is the standard measure for the elastic energy dissipation and termed internal friction or, less commonly, reciprocal quality factor.

The second factor, $(\Delta P)^2/4kT$, corresponds to the paraelastic polarizability, α, in the present simplified two-level model. In fact, one can show [12, 13] that for each of the two shear deformation modes in cubic crystals the internal friction can be expressed as

$$Q^{-1} = \Delta_M \cdot \frac{\omega\tau}{1+(\omega\tau)^2} \tag{3.13}$$

with a relaxation strength

$$\Delta_M = \frac{\varrho_0}{M} \cdot \alpha$$

where α is one of the polarizabilities $\alpha^{[\nu]}$ defined in Sect. 2.1.3. Thus, measurements of the elastic energy dissipation provide basically the same microscopic information as the quasi-static experiments discussed in Sect. 3.1.

Many different techniques may be employed to perform such measurements, as for instance ultrasonic attenuation, phase lag of forced vibrations, full width at half maximum of the resonance peak of forced vibrations, or damping of freely decaying vibrations. Many examples of the latter technique will be discussed later, therefore this technique will be outlined here in more detail. Damping of a freely decaying vibration can be expressed as

$$\varepsilon(t) = \varepsilon_0 \exp\left(-\frac{\omega}{2}Q^{-1}t\right)\exp(i\omega' t) \tag{3.14}$$

provided that $\Delta \ll 1$ [20]. Q^{-1} is as given by (3.13) and the oscillation frequency, ω', is given by

$$(\omega')^2 = \omega^2\left[1 - \frac{\Delta_M}{1+(\omega\tau)^2}\right] \tag{3.15}$$

where $\omega^2 = M_U/\Theta$ and Θ is an appropriate inertial moment.

$\omega'(\omega\tau)$ is related to Q^{-1} in terms of Kramers-Kronig relations, i.e. it describes the dispersive part of the response function. If $\omega\tau \ll 1$, the equilibrium alignment of the defects is achieved so that

$$(\omega')^2 = M_R/\Theta = M_U(1-\Delta_M)/\Theta \quad \text{if} \quad \omega\tau \ll 1 .$$

For $\omega\tau \gg 1$, where the dipole reorientation is frozen,

$$(\omega')^2 = \omega^2 = M_U/\Theta \quad \text{if} \quad \omega\tau \gg 1 .$$

From the difference between these two limiting frequencies one may obtain the relaxation strength, Δ_M. The difference between ω' and ω is small as long as the relaxation strength, Δ_M, is small. The shape of the functions $(\omega'(\omega\tau))^2$ and $Q^{-1}(\omega\tau)$ are of the same form as those of the in- and out-of-phase components shown in Figs. 3.2 and 3.3, as can be seen from a comparison of (3.8) and (3.13) and (3.15).

If the dipole reorientation obeys an Arrhenius type behavior (2.37)

$$\tau = \tau_0 \exp(E^R/kT)$$

the $\log \omega\tau$ scale is, for constant frequency, equivalent to an inverse temperature scale. In this case one has also to pay attention to the T^{-1} dependence of the relaxation strength so that only the functions $(Q^{-1}T)$ and $(\omega^2 - \omega'^2) \cdot T$ are symmetrical and antisymmetrical with respect to $\omega\tau = 1$, respectively. In particular, the T^{-1} dependence of $(\omega')^2$ may be taken as an experimental check of a truly paraelastic response obeying a linearized Boltzmann distribution with respect to site occupations.

3.3 Experimental Devices

As representative members of the many possible choices we will discuss here two devices, namely the so-called inverted torsion pendulum or Ke type pendulum, and the vibrating reed system.

The principle of the torsion pendulum is shown in Fig. 3.4. The sample is fixed at one end and clamped to a freely rotatable rigid inertia member at the other end. The sample may be twisted by means of an electromagnetic or electrostatic drive. The angular displacement, $\delta\varphi$, which is proportional to the torsional strain, is measured by some suitable pick-up system.

Figure 3.5 shows a practical construction, designed for low temperature operations $(5\,\mathrm{K} < T < 400\,\mathrm{K})$ and in situ low temperature $(T = 4.5\,\mathrm{K})$

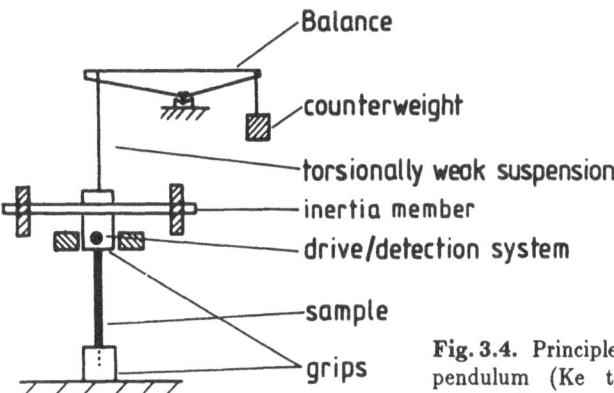

Fig. 3.4. Principle of an inverted torsion pendulum (Ke type pendulum). Further information in text

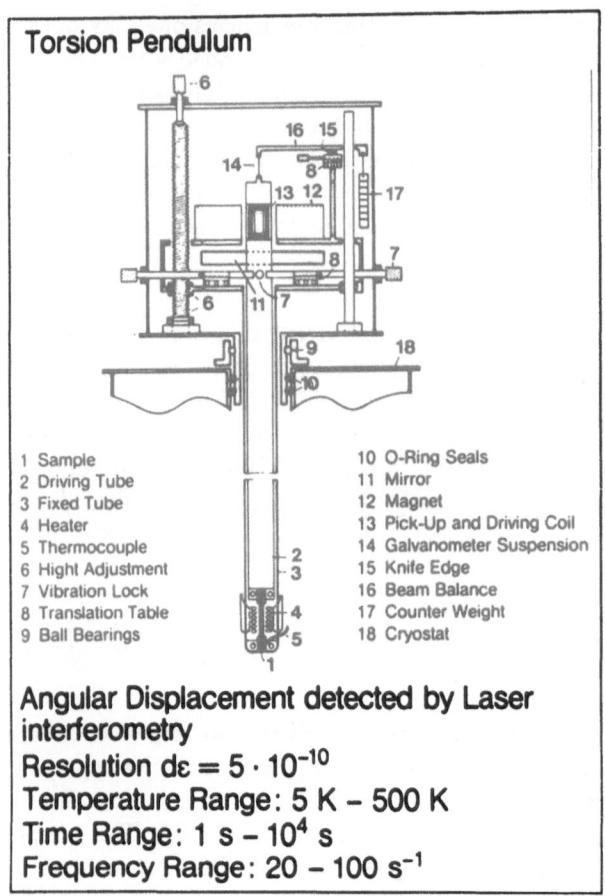

Fig. 3.5. Outline of a practical construction of an inverted torsion pendulum for operation and in situ electron irradiation at low temperatures ($5\,\mathrm{K} < T < 400\,\mathrm{K}$)

electron irradiations. The sample is positioned in the cold part of the central tube of a liquid helium cryostat. The connection between the sample clamps and the rigid frame and the free inertial member is provided by two thin-walled concentrical stainless steel tubes. The inertial moment may be changed by adding or subtracting lead weights to and from the drive tube.

The total weight is balanced by a counterweight so that the sample is kept free from longitudinal elastic loads. The angular deflection can be excited by a coil and magnet system, where the coil is fixed with respect to the inertia member, and the magnet with respect to the frame. Quasi-static experiments are performed by applying a constant electric current to the coil during a time period $0 \leq t \leq t_0$, as explained above. The angular deflection is determined by means of a laser interferometer capable of an angular resolution of $\delta\varphi = 10^{-8}$ rad. This instrument is described in detail in [17].

Torsional vibrations of the pendulum are excited by a phase shifted electric feedback loop. When the regulated feedback is switched off, the free decay

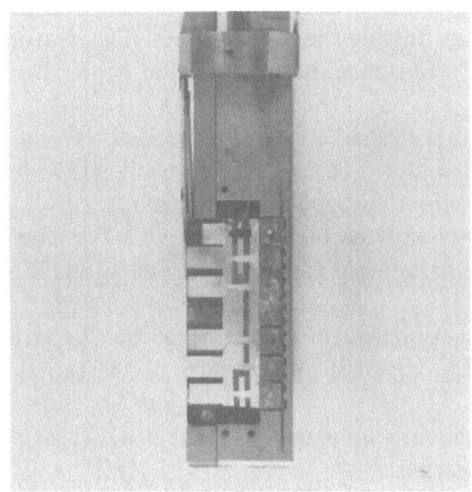

Fig. 3.6. Vibrating reed system, employing six samples simultaneously. The samples, which are tested sequentially, may be driven in flexure or torsion. One pick-up and drive electrode is visible at the position of the fourth sample from below. The shorter seventh sample is a dummy and used as a carrier for a thermocouple. The metal strip with the double ends behind is a resistivity sample used to determine defect concentrations in the case of irradiation experiments

of the oscillation occurs. It is measured by a pick-up coil which is in parallel to the drive coil.

As samples, any kind of cylindrical rods may be used. In the case of the irradiation experiments discussed later, tubular samples with cylindrical cross-section were used. Their diameter was typically $D = 3$ mm and the wall thickness $t = 0.1$ mm. For this sample geometry the torsional strain, ε, is related to the angular displacement, φ, by

$$\varepsilon = \frac{D}{2l} \cdot \varphi$$

where l is the free sample length.

In the vibrating reed device the sample has the form of a small, thin cantilever which is clamped at one end and free to vibrate at the other. A practical construction is shown in Fig. 3.6. This is a recent development of a multiple sample holder. The samples are shaped like "dog-bones". This shape has several advantages. First, during flexural deformation the strain is mainly concentrated in the narrow part. In this manner the strain at the clamp edge is reduced as compared to a sample with uniform cross-section. This reduces parasitic friction effects from the clamps. Second, due to the trapezoidal shape of the narrow part, the strain is constant within 10 % in this sample region [21, 22]. A further and particularly valuable advantage arises in irradiation experiments. As the strain is constant and concentrated in the narrow part, the exact positioning of the beam aperture with respect to the sample is not very critical, as long as its limits stay outside the narrow parts. This feature is important if absolute values of, for instance, internal friction effects from different irradiations are compared.

The samples are driven into either flexure or torsional oscillations via a fixed electrode behind each of the samples. The feedback drive signal can be tuned to either flexure or torsional vibration by electronic filtering.

The fixed electrodes serve additionally as pick-up electrodes. The variation of the capacity between sample and electrode is detected by an FM technique.

In principle also quasi-static experiments can be performed by this technique, since the frequency of the FM signal is proportional to the sample-electrode distance.

The present vibrating reed device can be placed into the same cryostat as the torsion pendulum described earlier.

During irradiation, both devices are positioned with respect to the cryostat such that the samples are located exactly between two irradiation windows, each of which consists of 24 μm thick stainless steels foils welded onto a steel frame. Fast electrons of 3 MeV kinetic energy may penetrate through these foils and through the samples, provided the sample thickness is kept small, typically 150 μm or less depending on the material. During irradiation at a temperature of about $T = 4.5$ K, liquid helium flows around the samples in the sample chamber, which is coupled into the flow circuit of a helium refrigerator system.

After irradiation, the cryostat is disconnected from that circuit and the samples cooled indirectly via a helium cooled-wall surrounding the measuring position on top of the irradiation position. In this mode, the helium is supplied from an internal helium reservoir. The samples may be transferred at the low temperature ($T \leq 5$ K) from the irradiation facility to the location where the measurements are performed. Here, the whole assembly is placed on a platform made of 3 tons of concrete suspended on elastic springs, which serves to decouple the apparatus from the acoustic noise of the surroundings.

During the measurements thermal coupling between the cold wall of the sample chamber and the sample is maintained by He-gas at a constant pressure of about 10 Pa. In order to eliminate apparent damping changes due to changes of the gas pressure, it is controlled at a constant value with an accuracy better than 0.01 Pa.

3.4   Use of Single Crystals

If single crystalline samples with specified orientations with respect to the torsional or bending axes are employed in the experiments, the measured relaxation strengths and relaxation times may be decomposed into the con-

Table 3.1. Relaxation strengths, Δ_G and Δ_E, of the effective shear- and Youngs moduli expressed in terms of the relaxation strengths Δ_C and $\Delta_{C'}$ for various sample orientations z. η is the anisotropy factor: $\eta = C/C'$

Cryst. orient.	<100>	<110>	<111>	Polycrystalline sample without texture
Torsion Δ_G	Δ_C	$\dfrac{1}{1+\eta}\Delta_C + \dfrac{1}{1+\eta^{-1}}\Delta_{C'}$	$\dfrac{1}{1+2\eta}\Delta_C + \dfrac{1}{1+0.5\eta^{-1}}\Delta_{C'}$	$0.6\Delta_C + 0.4\Delta_{C'}$
Bending Δ_E	$\dfrac{\frac{\Delta K}{3K} + \frac{\Delta C}{C}}{\frac{1}{3K} + \frac{1}{C}}$	$\dfrac{\frac{\Delta K}{9K} + \frac{\Delta C'}{12C'} + \frac{\Delta C}{4C}}{\frac{1}{9K} + \frac{1}{12C'} + \frac{1}{4C}}$	$\dfrac{\frac{\Delta K}{3K} + \frac{\Delta C'}{C'}}{\frac{1}{3K} + \frac{1}{C'}}$	$\dfrac{\Delta G}{1+\alpha} + \dfrac{\Delta K}{3(1+\alpha^{-1})}$

where:

$$\frac{1}{G} = \frac{3}{5}\frac{1}{C} + \frac{2}{5}\frac{1}{C'}$$

and

$$\alpha = \frac{G}{3K}$$

tributions from the deformation $\overset{[2]}{\varepsilon}$ to $\overset{[6]}{\varepsilon}$ in a unique manner. A detailed discussion of this point may be found, for instance, in [12]. The relations, which are of interest for the following chapters, are summarized in Table 3.1. As an example, a sample with a <100> torsional axis exhibits a pure mode [4] response corresponding to a change of the elastic modulus C alone. If the same sample were bent perpendicular to the <100> axis, the response would be a combination of compression (K) and <100> type shear (C'). The paraelastic change of the compressional modulus is zero (2.20), however, so that the paraelastic response is purely of <100> shear type. This is not so for the diaelastic effects, where ΔK is not negligible, in general.

4. Experimental Results for Pure Metals

4.1 Diaelastic Polarizabilities

The change of the elastic moduli ΔG or ΔE is directly related to the change of the square of the eigenfrequency, f, of the torsion pendulum or of the vibrating reed, respectively:

$$\frac{\Delta M_{\mathrm{d}}}{M} = \frac{\delta(f^2)}{f^2} \quad \text{with} \quad M = E \quad \text{or} \quad G . \tag{4.1}$$

E and G are the Youngs- and shear modulus, respectively. Their relaxation strengths are defined in Table 3.1 in terms of the relaxation strengths of K, C', C (2.19). The Frenkel pair concentration, ϱ_{FP}, is determined from the change of the residual electrical resistivity, Δr, of a special resistivity sample irradiated simultaneously with the bending or torsion sample.

The change of the resistivity is related to the concentration, ϱ_{FP}, by

$$\Delta r = r_{\mathrm{FP}} \cdot \Omega \cdot \varrho_{\mathrm{FP}} = r_{\mathrm{FP}} \cdot c_{\mathrm{FP}} . \tag{4.2}$$

A recent collection of specific resistivity data r_{FP} may be found in [2].

The diaelastic polarizabilities may then be determined from the slope of ΔM_{d} with respect to Δr as

$$\alpha_{\mathrm{d}} = -\frac{\Delta M_{\mathrm{d}}}{\varrho_{\mathrm{FP}}} = -M r_{\mathrm{FP}} \Omega S \tag{4.3}$$

where S is the slope of $\delta(f^2)/f^2$ with respect to Δr.

In Fig. 4.1 results are shown for single crystalline samples of Al subjected to electron irradiation (3 MeV) at 4.7 K. The relative modulus change expressed in terms of $\delta(f^2)/f^2$ was measured in a torsion pendulum. After an initial increase the expected linear slope of the frequency is observed with increasing irradiation dose.

The initial increase is due to a dislocation pinning effect [12]. Dislocations in general bow out under the influence of an external stress. This gives rise to a strain in addition to the purely elastic strain, which is equivalent to a reduction of the appropriate elastic modulus. Point defects like self-interstitials and vacancies act as pinning centers for the dislocations so that they gradually inhibit the dislocation motion and thereby increase the elastic modulus

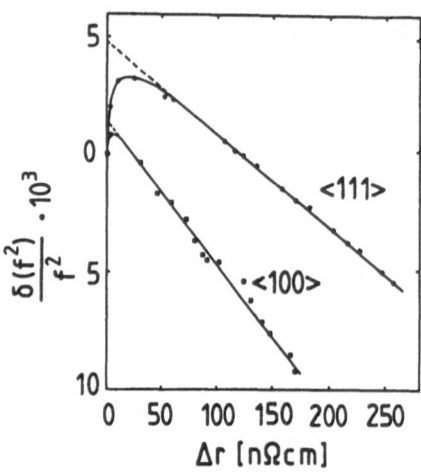

Fig. 4.1. Relative change $\delta(f^2)/f^2$ of the torsional frequencies f of single crystalline aluminum samples as a function of the irradiation-induced resistivity change, Δr. The torsional axes were <100> and <111> oriented. The irradiations with 3 MeV electrons and the measurements of the frequencies and resistivities were performed at 4.5 K

of the material with increasing defect concentration. Figure 4.1 demonstrates that this pinning process is very effective: a concentration of about 50 atppm immobile Frenkel pairs was sufficient to remove any observable dislocation effect in the present case.

The diaelastic modulus change is not the sole source of the linear decrease of the frequency with dose. One further contribution arises from the geometrical change of the sample shape due to the lattice expansion induced by the Frenkel pairs. Since this expansion is well known from lattice parameter measurements this contribution is readily calculated as [19, 25]:

$$\frac{\delta(f^2)}{f^2}\bigg|_{\text{expans.}} = -1.9 c_{\text{FP}} \tag{4.4}$$

for the sample geometry employed in these particular experiments. From Table 3.1 and the slopes of the linear parts in Fig. 4.1, one obtains for the relative changes of C and C':

$$\frac{1}{c_{\text{FP}}} \cdot \frac{\Delta C}{C} = -27 \pm 2$$

$$\frac{1}{c_{\text{FP}}} \cdot \frac{\Delta C'}{C'} = -16 \pm 2 \,.$$

In order to compare the experimental data with theoretical data, a correction term due to anharmonic effects has to be introduced. This is because of the difference in the lattice expansion of a crystal of finite size, such as the sample as compared to an infinite crystal, for which the theoretical values have been calculated. The anharmonic change of the elastic moduli due to the additional lattice expansion may be expressed by the pressure derivatives of the elastic constants as [11]

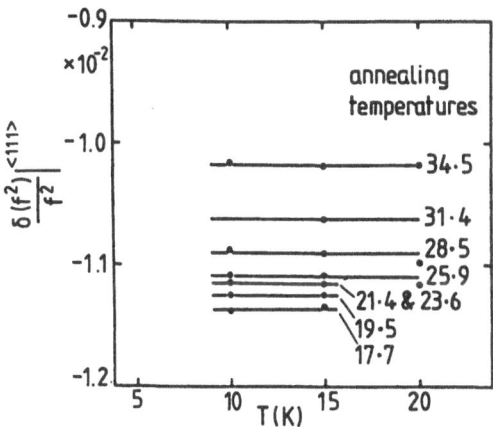

Fig. 4.2. Temperature dependence of the frequency change $\delta(f^2)/f^2$ of the <111>-oriented Al sample after electron irradiation and after 10 min anneals at the indicated temperatures

$$\left[\frac{1}{c}\cdot\frac{\Delta C}{C}\right]_{\text{anh}} = K\cdot\frac{\delta V}{\Omega}\cdot\frac{1}{C}\cdot\frac{\partial C}{\partial p}$$

$$\left[\frac{1}{c}\cdot\frac{\Delta C'}{C'}\right]_{\text{anh}} = K\cdot\frac{\delta V}{\Omega}\cdot\frac{1}{C'}\cdot\frac{\partial C'}{\partial p}$$

(4.5)

where the volume difference δV is, for isotropic materials, given by

$$\delta V = V^{\text{rel}}\frac{2(1-2\nu)}{3(1-\nu)}$$

with $\nu =$ Poissons's number.

For the present case of Al, the above corrections amount to -3.6 for C and -3.2 for C' [25]. With these one finds finally $\overset{[4]}{\alpha}_{\text{d}} = 269\,\text{eV}$ and $\overset{[2]}{\alpha}_{\text{d}} = 125\,\text{eV}$.

An analysis of the temperature dependence proves that this modulus decrease is purely diaelastic and does not contain any paraelastic effects proportional to $1/T$. This is shown in Fig. 4.2 where the relative change of the torsional modulus observed after irradiation and after 10 min anneals at the indicated temperatures is shown as a function of temperature for a <111>-oriented sample. No indication of a $1/T$ dependence indicating a reorientation process is evident.

The polarizabilities contain the contribution of the self-interstitials and of the vacancies. In order to extract the self-interstitial effect alone, one needs polarizabilities of the vacancies. Reliable experimental values are not available. One is left with the possibility of using the theoretical values obtained from the computer simulation Table 2.3.

For Cu in a similar experiment values of $\overset{[4]}{\alpha}_{\text{d}} = 377\,\text{eV}$ and $\overset{[2]}{\alpha}_{\text{d}} = 111\,\text{eV}$ were found [26]. In an independent experiment, using an ultrasonic technique the same values were found for Cu irradiated with thermal neutrons [16, 27].

49

A comparison of the experimental Al and Cu data with the theoretical results listed in Table 2.3 demonstrates the close agreement between theory and experiments. The anisotropy of the diaelastic polarizabilities, $\alpha_d^{[4]} > \alpha_d^{[2]}$ is compatible with a tetragonal structure such as that of the <100> split configuration.

From measurements on Mo it was found that the opposite relation holds for bcc metals, $\alpha_d^{[4]} < \alpha_d^{[2]}$, namely $\alpha_d^{[4]} = 162\,\mathrm{eV}$ and $\alpha_d^{[2]} = 298\,\mathrm{eV}$ [18]. This gives evidence for the existence of resonance modes in the bcc structure as well. In Fig. 4.16 the arrangement of springs and the displacement pattern of resonant modes for a <110> dumbbell in a bcc lattice is shown. This figure is again a result of the computer simulations and will be discussed in more detail in a later chapter. Here it is referred to in order to show the analogy to the resonant modes of the <100> dumbbell in the fcc lattice.

However, the coupling of the springs to the eigendeformation $\varepsilon^{[\nu]}$ is not so simple in the bcc case as in the fcc case. For instance, the central dumbbell spring can be coupled via vibration modes I and II to the C and C' type deformations, respectively. From this qualitative consideration alone a forecast of the anisotropy is not possible. Computer simulations so far have only dealt with Fe potentials only for the bcc structure [133]. They indicate much smaller polarizabilities for the self-interstitial, and a much less pronounced anisotropy for both, the self-interstitial alone and for the Frenkel pair. From the much smaller migration energy of the self-interstitial in Mo as compared to Fe (see the following Chap. 5) one expects larger polarizabilities in Mo compared to Fe in agreement with the present experimental findings. A theoretical support for the anisotropy observed in Mo is not available.

The recovery of the polarizabilities and of the residual electrical resistivity of the Al samples are shown in Fig. 4.3. The data have been normalized to the values obtained after completion of the irradiation. The measurement temperature was about 5 K in all cases. The bottom figures of Fig. 4.3 show the average polarizability of the defects present in the sample at any annealing temperature, i. e. the ratio of

$$(\delta(f^2)/f^2)/\Delta r \sim \sum_i \varrho_i \alpha_d^i \Big/ \sum \varrho_i r_i$$

where the sum (i) extends over all kinds of defects present in the sample at that temperature.

The large resistivity recovery between 20 K and 38 K is associated with the long-range diffusion of the self-interstitials and their subsequent recombination with the vacancies and/or their clustering with each other. Between 20 K and 36 K the modulus recovery, $\delta(f^2)/f^2 \Delta r$, roughly parallels the production curves of Fig. 4.1. This indicates that the process is mainly the inverse of the production process, i. e. recombination is the predominant reaction going on in this temperature regime. Between 36 K and 43 K the modulus

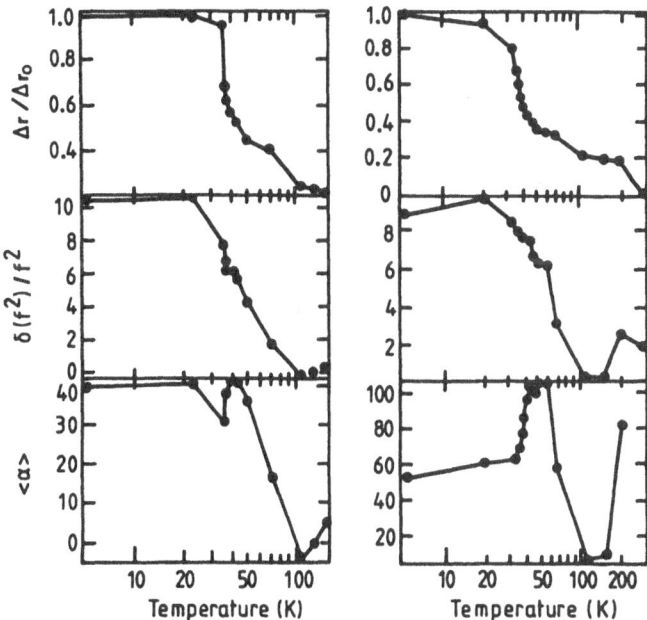

Fig. 4.3. Annealing behavior of the normalized resistivity change (*top*), change of vibrational frequency of $\delta(f^2)/f^2$ (*middle*) and the average polarizability (*bottom*) for a <111>-oriented (*left*) and a <100>-oriented (*right*) Al crystal after 3 MeV electron irradiation to $\Delta r_0 = 270\,\text{n}\Omega\,\text{cm}$ and $170\,\text{n}\Omega\,\text{cm}$, respectively. The average polarizability is given by the ratio of the two upper curves and further discussed in the text

decreases much less than the residual resistivity, i. e. the average polarizability per self-interstitial increases by almost a factor of 2 (bottom figure). This increase may be attributed to the formation of small self-interstitial clusters at the end of stage I, i. e. 40 K. It indicates that di-interstitials and probably also tri-interstitials have an even higher diaelastic polarizability than single self-interstitials in Al.

In the temperature regime $43\,\text{K} < T < 110\,\text{K}$ the resistivity changes only slightly. This indicates that only little recombination occurs, and in fact from Huang scattering experiments [28] the predominant self-interstitial reaction was found to be a growth of the interstitial clusters to an average size of 10. This cluster growth is accompanied by a dramatic decrease of the diaelastic polarizability. Obviously, self-interstitials in larger clusters lose their high polarizability. This temperature regime coincides further with the temperatures above which interstitial-related relaxation effects have vanished as discussed later. The modulus recovery occurring above 110 K involves additional dislocation reactions which we will not discuss further here.

The modulus change observed for the <111> crystal, which contains $\overset{[2]}{\alpha}$ contributions in addition to the $\overset{[4]}{\alpha}$ contribution of the <100> sample, shows the same general trend. However, there is a quantitative difference in that the increase of the polarizability at the end of stage I is much less pronounced than for the <100> sample. This means that the $\overset{[2]}{\alpha}$- polarizability does not undergo any dramatic change when small self-interstitial clusters are formed.

The existence of resonant vibrations of self-interstitials and their relation to the diaelastic polarizabilities has been corroborated by measurements of two other effects. First, *Holder* et al. [16] looked more closely at the small temperature dependence of the elastic moduli before and after annealing in stage I of Cu. They found a temperature dependence which is compatible with the assumption of an additional line in the phonon spectrum. They estimated its frequency as about $0.8 \times 10^{12}\,\text{s}^{-1}$, which is rather close to the value expected from theory for the resonant modes. This temperature dependence is so small that it would be not visible in Fig. 4.2.

The second kind of experiments are inelastic neutron scattering studies, which provide direct information on resonant modes present in the phonon spectrum. *Nicklow* et al. [29] found a characteristic shift of the phonon line with the scattering vector in neutron-irradiated Cu, which was consistent with the type of resonant perturbation anticipated for the <100> dumbbell self-interstitial. The slope of the dispersion line at small scattering vectors was found to be compatible with the polarizabilities given above, and from the cross-over point with the undistorted dispersion line, a resonant mode frequency of $0.8 \times 10^{12}\,\text{s}^{-1}$ was concluded.

In a very recent experiment diffuse inelastic neutron scattering measurements were performed on electron-irradiated Al single crystals at scattering vectors q far off the phonon lines [134]. So-called constant-q scans were performed in an energy range from 0 to 2 THz. A spectrum consisting of a pronounced line at 0.7 THz and a broad superposition of several lines above 1.4 THz was observed. As the most probable explanation it was suggested, that the line at 0.7 THz is caused by the A_{2u} mode and the shoulder at 1.4 THz by the E_g mode of the single self-interstitial. This assignment leads in particular to a good agreement with the previous diaelastic polarizability data, as well as with the migrational energy of the self-interstitial to be discussed in the next chapter.

In summarizing this chapter we may conclude the following:

1) Interstitial defects elastically soften the crystal. The corresponding diaelastic polarizabilities of single self-interstitials are large and anisotropic.

2) The anisotropies observed in Al, Cu and Mo are consistent with the symmetry of the <100> split configuration in fcc metals and of the <110> split configuration in bcc metals.

3) The anisotropy and the magnitude provide strong experimental evidence for the existence of low-frequency resonant vibrations of self-interstitials. These resonant modes are in turn the physical basis for understanding the high mobilities of self-interstitials observed in the stage I recovery experiments.

4) Small self-interstitial atom clusters ($n \approx 2-3$) possess a high polarizability, $\overset{[4]}{\alpha}$, whereas larger clusters ($n \geq 10$) do not. Therefore high mobilities

of self-interstitials in small clusters may be expected while larger-sized clusters will have a low mobility.

4.2 Diffusional Motion of Self-Interstitials

4.2.1 fcc Metals

a) Aluminum

The first elastic aftereffect experiment devoted to studying the mechanism of self-interstitial migration was performed on Al [31, 32] using the equipment described in Sect. 3.4. The single crystalline Al samples, hollow cylinders with <100> and <111> torsion axes, received irradiations with 3 MeV electrons leading to maximum Frenkel pair-concentrations of 675 and 425 atppm, respectively. Examples of isochronal aftereffects curves are shown in Fig. 4.4 for a <111> sample and in Fig. 4.5 for a <100> sample after various steps in isochronal annealing procedures. Figure 4.6 shows results of an internal friction experiment performed on an electron irradiated polycrystalline Al

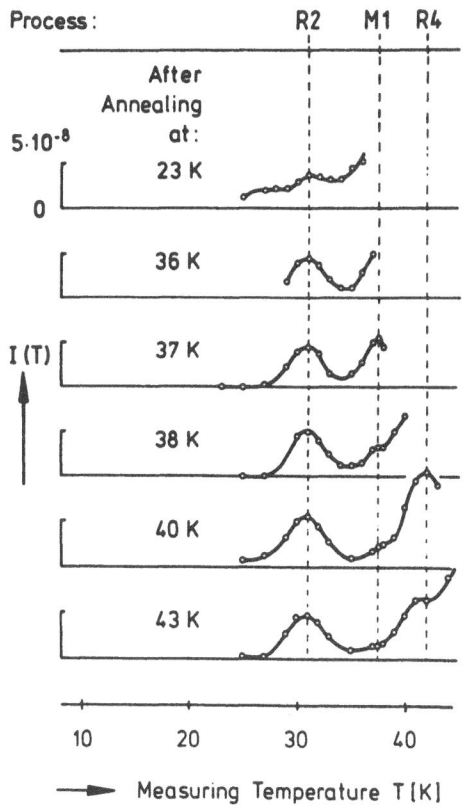

Fig. 4.4. Isochromal elastic aftereffect curves obtained in an <111>-oriented Al sample containing Frenkel pairs with concentrations $c_{FD} = 675$ atppm. The temperatures in the diagram are the restrictive annealing temperatures which are identical to the maximum attained temperature of the previous curve

53

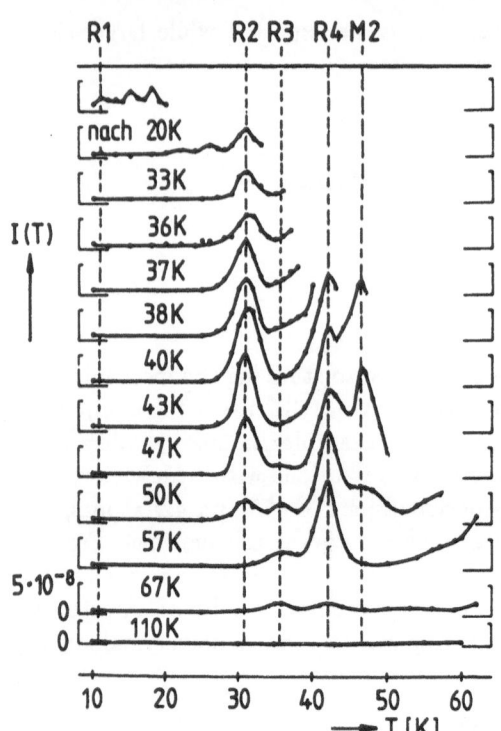

Fig. 4.5. Similar to Fig. 4.4, but a
<100>-oriented sample containing
$c_{FD} = 425$ atppm Frenkel pairs

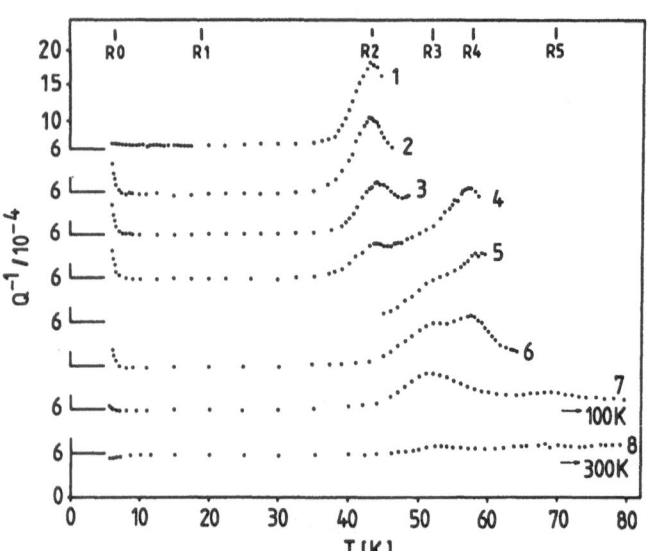

Fig. 4.6. Internal friction spectrum of an electron-irradiated polycrystalline Al sample
containing $c_{FD} = 500$ atppm Frenkel pairs

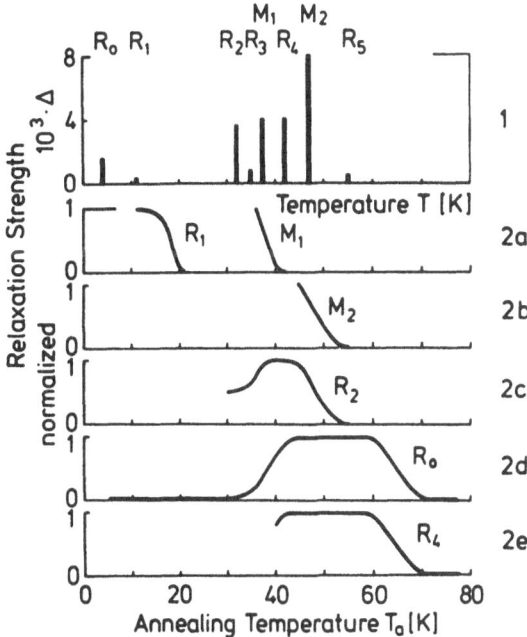

Fig. 4.7. Schematic representation of the relaxation spectrum at $\tau = 100\,\mathrm{s}$ of electron-irradiated Al (*top, 1*) and of the annealing curves of the different processes (*bottom, 2a to 2e*). Further explanations are given in the text

sample. These relaxation spectra exhibit a total of 8 different maxima labelled $M1$, $M2$ and $R0$ to $R5$. Label M serves to indicate a relaxation process involving concomitant long-range defect migration, whereas R refers to a localized reorientation. M processes do not show up in the internal friction spectra. The reason for this is that the large number of migrational jumps required with the detection of a peak in the internal friction measurement leads to defect annihilation before the measurement is completed. Therefore, application of the elastic aftereffect technique was essential in order to detect process $M1$ and $M2$.

Another reason for the absence of a relaxation process may be the orientation dependence of the effect. An example of this kind is again provided by process $M1$, which is visible in the spectrum of the $\langle 111 \rangle$ sample, but not in that of the $\langle 100 \rangle$ sample.

An overview over the whole relaxation spectrum and the annealing behavior of the various processes is displayed in Fig. 4.7. In the upper part, the relaxation processes are represented as lines located at temperatures where their respective relaxation times are $100\,\mathrm{s}$. The heights are as observed in the $\langle 100 \rangle$ specimen, except for process $M1$, which has been taken from the $\langle 111 \rangle$ specimen and $R0$, which has been taken from the internal friction results. The lower part of Fig. 4.7 shows the behavior of the relaxation lines as a function of the annealing temperature during an isochronal annealing pro-

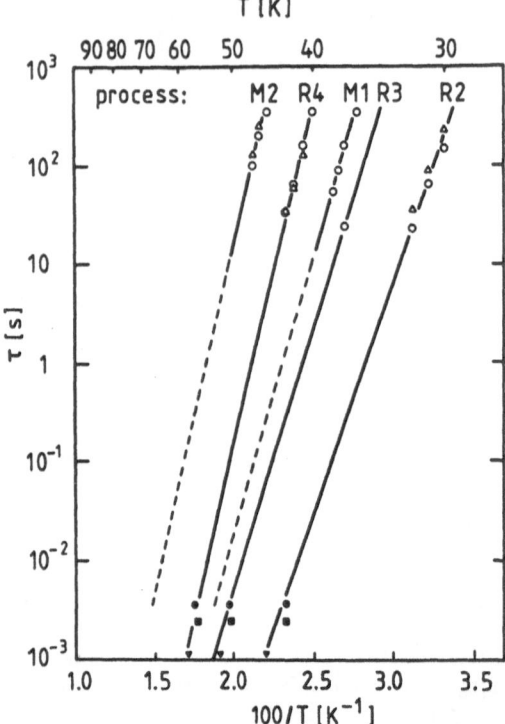

Fig. 4.8. Arrhenius diagram $\tau = \tau(T^{-1})$ displaying the temperature dependence of the relaxation time of the different relaxation processes observed in electron-irradiated Al. The solid lines through the data points of processes $M1$ and $M2$ change into dashed lines at those temperatures where their relaxation strength has dropped to zero due to annealing. Explanation of symbols: o — elastic aftereffect data of a <111>-oriented sample, $\triangle$ — elastic aftereffect data of a <100>-oriented sample, • — internal friction data of a polycrystalline sample. ■, ▼ — internal friction data of α- and neutron-irradiated polycrystalline Al [33, 34]

gram. All data were normalized to their respective maximum value attained during the annealing procedure. Since the relaxation strength is proportional to the defect concentration, these curves directly give the change of the relative concentrations c/c_{max} of the individual types of defects responsible for the respective processes.

In the following we will discuss the various processes separately as they are attributed to different atomic self-interstitial models and jump modes.

To complete the experimental findings, Fig. 4.8 shows the Arrhenius plot for all processes except for $R0$ and $R5$, for which they could not be determined unambiguously because of the low temperature position of $R0$ and because of the small magnitude of $R5$. For comparison, internal friction data from other sources [33, 34] have been included. The full lines through the data points of processes $M1$ and $M2$ end at those temperatures where according to Fig. 4.7, their relaxation strength has fallen to zero.

Process M1. Of particular interest in the present discussion is peak $M1$, which appears in Fig. 4.4 at 37.5 K. The behavior of peak $M1$ after different annealing temperatures indicates that the defect responsible for this anelastic process anneals in the same temperature regime where the peak occurs. In order to show the annealing behavior of processes $M1$ in greater detail, the relaxation strength observed at a reference temperature of 37 K is plotted in Fig. 4.9 as a function of the residual resistivity measured at 6 K after each annealing temperature. Process $M1$ anneals completely in stage I_{D+E} defined by the recovery range $(0.7 > \Delta r/\Delta r_0 > 0.5)$. The annealing of process $M1$ parallels the resistivity recovery in stage I_{D+E}, indicating that its relaxation strength is proportional to the concentration of single self-interstitials present in the sample. Therefore, process $M1$ is attributed to the anelastic relaxation of single self-interstitials.

It is apparent from Fig. 4.9a that a small portion of the relaxation strength measured at 37 K is due to defects which do not anneal until the middle of stage II. However, the time constant (50 s) for this background at 37 K is quite different from that of process $M1$ (160 s) (see Fig. 4.10). Part of this background effect is due to the two neighboring processes R_2 and R_4 but other effects apparently also contribute but cannot be individually separated because of the small magnitude of the total background. The annealing of the relaxation strength in the <100>-oriented crystal measured at the reference temperature of 37 K is shown in Fig. 4.9b. No indication of a process which anneals systematically in stage I_{D+E} is found in the <100> sample. How-

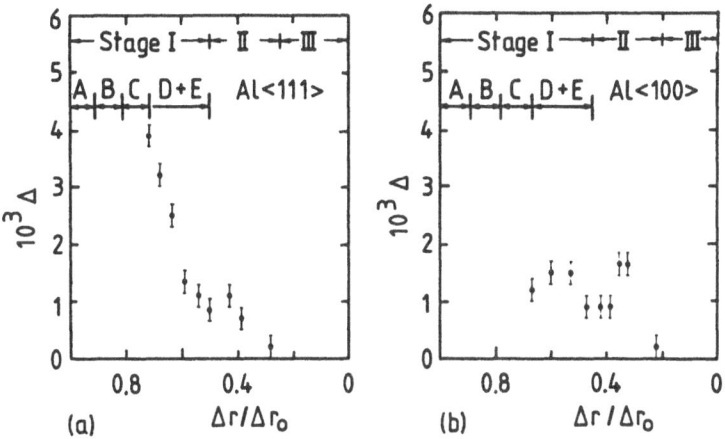

(a)   (b)

Fig. 4.9a,b. Relaxation strength of process $M1$ as a function of the annealing treatment expressed in terms of the residual resistivity $\Delta r/\Delta r_0$. The respective left-hand sides correspond to the irradiated state, the right-hand sides to the fully annealed state. Annealing of $M1$ is observed at the end of stage I. There is no contribution of the $M1$ process visible in the <100>-oriented sample (b), $(\Delta r_0 = 170\,\text{n}\Omega\,\text{cm})$, but only in the <111>-oriented one (a), $(\Delta r_0 = 270\,\text{n}\Omega\,\text{cm})$. The effect is superimposed on a background apparent in both orientations

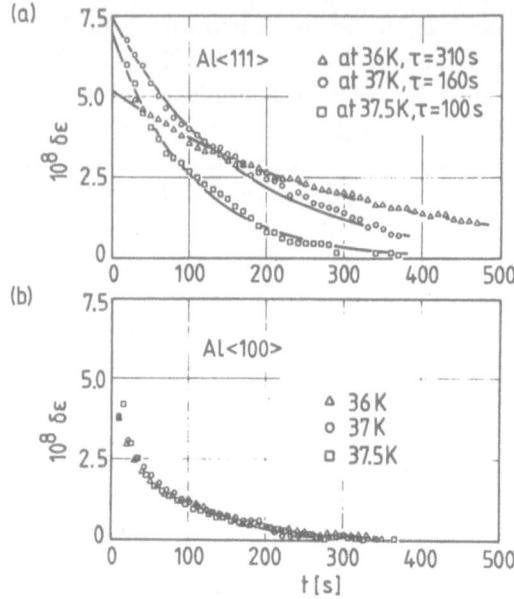

Fig. 4.10. (a) Elastic aftereffect curves exhibiting relaxation process $M1$ (<111> orientation) at the three different temperatures indicated. (b) The <100> orientation shows only the temperature-independent background effect

ever, a background similar to that found in the <111> orientation is again present. The difference between process $M1$ and the background effect shown in Fig. 4.9 also manifests itself in the temperature dependence of the observed relaxation effect. Figure 4.10 shows the results obtained at 36, 37 and 37.5 K. As expected for an Arrhenius type relaxation process, a strong temperature dependence is found for the relaxation rate extracted for the <111> sample. On the other hand, no significant temperature dependence of the time constant is observed for the <100> sample which does not exhibit process $M1$ but only the background process. In other words, there is no detectable relaxation process from single interstitials in the <100>-oriented crystal.

From the absence of the relaxation effect in the <100> crystal in Al, it follows directly that the self-interstitial must have either tetragonal or <100> orthorhombic symmetry. Of the several proposed configurations for the self-interstitial in fcc metals only the <100> split configuration is consistent with this symmetry restriction. This result is in agreement with the results of diffuse x-ray work discussed below and with the diaelastic modulus measurements discussed above.

If the concentration of defects is known, the anisotropy $\overset{[2]}{\alpha}$ of the dipole tensor characterizing the tetragonal interstitial can be calculated from the relaxation strength of the $M1$ process according to (2.20).

However, caution must be exercised in using the measured resistivity changes to determine the change in the concentration of single self-interstitials.

Only single interstitials contribute to the mechanical relaxation process $M1$ but the resistivity is not as selective. The amount of resistivity remaining in the sample is a measure of the total defect concentration and contains a contribution from all the clusters which may have already formed during irradiation or annealing. Thus by using the amounts of relaxation strength and resistivity which remain, a lower limit for $|P_{11} - P_{22}|$ can be calculated (2.20b), while the combination of annealed relaxation strength and annealed resistivity produces an upper bound. In this manner, using a resistivity change of $4\,\mu\Omega\,\text{cm/at\%}$ defects [2], a value of

$$|P_{11} - P_{22}| = (1.1 \pm 0.3)\,\text{eV}$$

is found. It is in good agreement with both the theoretically calculated values (Table 2.2) and the values determined from x-ray measurements [42].

The magnitude of $|P_{11} - P_{22}|$ for the <100> split interstitial is relatively small, as compared for example with a value of $6.0\,\text{eV}$ for carbon in iron. This indicates that the long-range part of the displacement field around the single self-interstitial has nearly cubic symmetry. If central forces are used to describe the interactions between the two <100> split dumbbell atoms and their nearest neighbors, a value of $|P_{11} - P_{22}| = 1.1\,\text{eV}$ is consistent with a value for the distance, d, between the two dumbbell atoms of either $0.4a$ or $0.6a$, where a is the lattice constant.

An activation analysis for process $M1$ is shown in Fig. 4.8, where the measured relaxation time, τ, is plotted as a function of the reciprocal measuring temperature $1/T$. The slope and intercept for the resulting straight line give values of

$$E = (0.115 \pm 0.025)\,\text{eV} \quad \text{and} \quad \tau_0 = 10^{-14 \pm 3}\,\text{s}$$

for the activation energy, E, of the reorientation and for its preexponential time constant, τ_0, respectively. This activation energy is the same as that reported for resistivity annealing in stage I_{D+E}, indicating that both reorientation and migration of the self-interstitial occur by the same jump process.

From the rate of decrease of the relaxation strength during each annealing step one can obtain an estimate of the average number of jumps, n, that a self-interstitial has made before reacting with another defect and disappearing. A value of $n \approx 10$ is found in the middle of stage I_{D+E}.

Together, all experimental findings discussed till now provide clear experimental evidence that the jump mechanism inherent in relaxation process $M1$ is the basic diffusion jump of the <100> split self-interstitial, namely the combined reorientation-translation jump shown in Fig. 2.12. The reorientation of the dumbbell axis is the source of relaxation process $M1$, while the shift of the dumbbell center by a nearest neighbor distance, $a/\sqrt{2}$, enables the three-dimensional long-range diffusion. The present experiments also show that reorientation without migration requires a higher activation energy, since no relaxation process due to single interstitials is observed at a temperature below that of free interstitial migration.

Process $M2$. Process $M2$ is centered at about $47\,\mathrm{K}$ in the elastic aftereffect results for both crystal orientations, but is absent from the internal friction data (Fig. 4.6). Like process $M1$, $M2$ also anneals in the same temperature interval where the relaxation occurs. Following the same reasoning as before this shows that defect reorientation occurs simultaneously with migration.

The temperature dependence of the relaxation time for process $M2$ yields (Fig. 4.8).

$$E = (0.135 \pm 0.25)\,\mathrm{eV} \quad \text{and} \quad \tau_0 = 10^{-13 \pm 3}\,\mathrm{s}$$

for the activation energy and preexponential time constant, respectively.

A comparison of the recovery of the relaxation strength of $M2$ (Fig. 4.7) with that of the resistivity (not shown here) reveals that the defect responsible for this process anneals at temperatures slightly above stage I, in the region called stage II_1. Huang scattering measurements [28] have shown that stage II_1 is primarily due to the disappearance of the di-interstitials present at the end of stage I. Therefore, process $M2$ is assigned to the simultaneous migration and reorientation of di-interstitials.

Assuming that the concentration of defects responsible for peak $M2$ is proportional to the initial defect concentration produced by the irradiation, the ratio of the relaxation strengths related to the same self-interstitial concentration, in the two different orientations is, see Table 3.1,

$$\Delta_G^{<100>}/\Delta_G^{<111>} = 3.7 \pm 0.4 \; .$$

This value indicates that the process $M2$ defect has lower than tetragonal symmetry, but not $<100>$ orthorhombic.

According to the computer simulation the di-self-interstitial consists of two parallel dumbbells whose axes are slightly tilted away from the $<100>$ direction (Fig. 2.16a). This model exhibits two different jump modes which give rise to a simultaneous reorientation and long-range migration as shown in Fig. 4.11.

The first jump produces a $90°$ rotation of the $<110>$ di-self-interstitial, and a two-dimensional migration in a (100) plane of the defect center-of-mass (upper row of Fig. 4.11). For the second mode (second row Fig. 4.11), one of the dumbbells passes into the same saddle point configuration as before but then a stable configuration is formed by a rotation by $90°$ of the other self-interstitial to stabilize the configuration. This produces a $60°$ rotation of the $<110>$ di-self-interstitial and a three-dimensional migration of the center of mass. The occurrence of a $90°$ and a $60°$ rotational jump is essentially characteristic of a $<110>$ orthorhombic defect, as evident from Table 2.1. However, the dipole force tensor of the present di-self-interstitial is not exactly $<110>$ orthorhombic [38]. As long as only $60°$ and $90°$ jumps are involved,

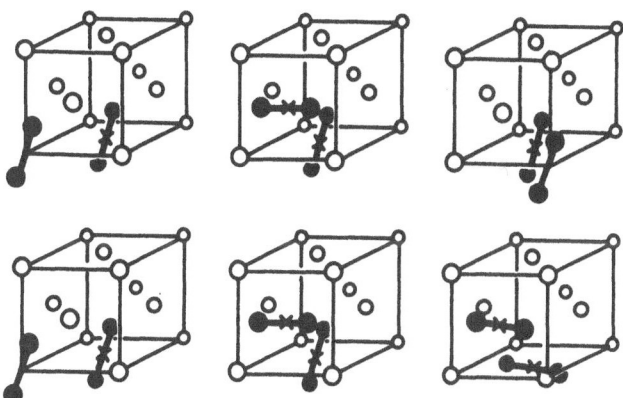

Fig. 4.11. Migration jump modes of the di-interstitial suggested by computer simulation. For a detailed explanation refer to text

however, one can discuss the relaxation phenomena in terms of a $<110>$ orthorhombic defect. The relaxation time constants τ_C and $\tau_{C'}$, appropriate for the two shear elastic constants C and C', are different. If ν_{ij} is the jump frequency for motion from configuration i into configuration j, the appropriate relaxation times are given by

$$1/\tau_{C'} = 6\nu_{13} \quad \text{and} \quad 1/\tau_C = 2\nu_{12} + 4\nu_{13} \ .$$

The existence of two different relaxation modes suggests the possibility of observing two separate relaxation peaks due to di-interstitials. However, the computer calculations suggest that the process $1 \rightarrow 2$ is much easier to excite than $1 \rightarrow 3$, i.e. the nearest neighbor jump is energetically more favorable than a pure rotation. Hence the $1 \rightarrow 3$ mode would occur at such a temperature that the vast majority of the di-interstitials would have already disappeared during migration by the $1 \rightarrow 2$ mode. It follows that only the relaxation process described by

$$1/\tau_C = 2\nu_{12}$$

should be observed experimentally. This leads to an expected ratio of the relaxation strengths for the di-interstitial in aluminum ($\eta = 1.2$) according to Tables 2.1 and 3.1, of

$$\Delta_G^{<100>}/\Delta_G^{<111>} = 3.4 \ .$$

With experimental error, this is the value found for process $M2$.

The assumption that just above stage I all the remaining residual resistivity change is due to di-interstitials and vacancies, yields an upper limit for the concentration of di-interstitials present in the sample. For the defect anisotropy from (2.20c), this gives a value of

$$\sqrt{P_{12}^2 + P_{13}^2 + P_{32}^2} \geq 6 \, \text{eV} \ .$$

This is in reasonable agreement with a theoretical value of 8.66 eV [38] and with a value of 6.7 eV as obtained from the Huang diffuse scattering [28].

It should be remarked here that the experimental findings can be explained equally well by the di-interstitial model proposed originally by *Johnson* [35] where the two dumbbells are exactly parallel to the <100> direction and the dipole tensor is exactly <110> orthorhombic [36]. In this case, $P_{13} = P_{32} = 0$, and $P_{12} \geq 6 \, \text{eV}$.

Process R2. We now turn to process R2 which appears in the internal friction results (Fig. 4.6) near 42 K, well above the onset of long-range interstitial migration. The high-temperature side of the corresponding internal friction peak is distorted by annealing (see Fig. 4.7). The information pertaining to process R2 derived from the internal friction measurements is therefore limited. In the elastic aftereffect measurements (Figs. 4.4 and 4.5), process R2 is found at 31 K. From Fig. 4.8, values of

$$E = (0.092 \pm 0.009) \, \text{eV} \quad \text{and} \quad \tau_0 = 10^{-13 \pm 1} \, \text{s}$$

are obtained for the activation energy and preexponential time constant, respectively.

The lower observation temperature of 31 K also removes the complication of simultaneous annealing, and permits the relaxation strength of the process to be studied during state I_{D+E} recovery. Process R2 begins to grow in strength in annealing stage I_{D+E}, and its increase is particularly sharp near the end of this stage (Fig. 4.7). A few of the process R2 defects are apparently created during irradiation, but the primary mechanism for their production involves the long-range migration of the self-interstitials in stage I_{D+E}. Process R2 disappears in a parallel fashion with process M2 after annealing in stage II_1.

Because of the similarity in recovery exhibited by processes R2 and M2, an obvious interpretation would be the assignment of process R2 to a non-migratory reorientation of the M2 defect. Such a separation of relaxation processes is not unknown, and is referred to in the literature as "frozen-free split", [12].

In principle the di-interstitial with slightly tilted dumbbells offers the possibility of such a process, since there are two equivalent tilt directions between which the two dumbbells may rock back and forth in a coupled manner (Fig. 4.12) [37]. This motion would give rise to an anelastic response in only the C mode, which is in agreement with the experimentally observed ratio of relaxation strengths of process R2, $\Delta_G^{<100>}/\Delta_G^{<111>} = 3.6 \pm 0.4$ [37]. However, the computer simulation yields an activation energy for the tilt-flip motion of 0.01 eV which is almost a factor of ten smaller than the 0.092 eV observed for process R2 [38].

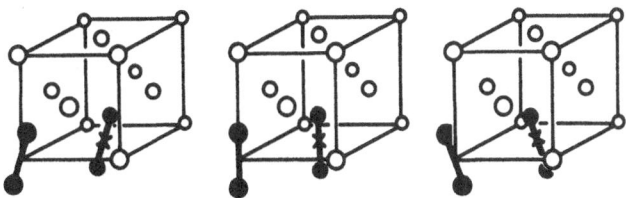

Fig. 4.12. "Tilt-flip" mode of the di-interstitial

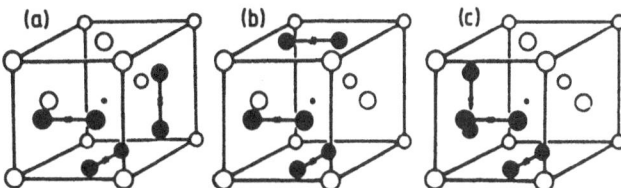

Fig. 4.13. The reorientation jump of a tri-interstitial (of Fig. 2.16) occurs by two subsequent combined jumps of one of the dumbbells, whereby the other two dumbbells keep their positions and orientations

From the view point of activation energies one would rather tend to assign process $R2$ to the reorientation of the tri-interstitial shown in Fig. 4.13 [36]. The motion of the tri-interstitial is also a local reorientation without concomitant long-range migration and its activation energy is slightly lower than the migration energy of the single self-interstitial [38]. The parallel annealing of processes $M2$ and $R2$ must then be attributed to a reaction in which both the di- and tri-interstitials are involved.

Process $R0$ and $R4$. Similar correlations in annealing behavior exist between two other processes, $R0$ and $R4$. Process $R0$ appears in the internal friction measurements near 6 K, the lowest attainable measuring temperature. However, its true maximum occurs at an even lower temperature. It is the most easily activated of all the observed processes, and since it is buried in the instantaneous part of the strain recovery, it does not appear in the isochronal aftereffect curves. Because of the failure to observe a true peak maximum, no accurate values can be given for its activation energy and preexponential time constant.

Process $R4$ appears in both the internal friction and elastic aftereffect results. The behavior of its relaxation strength during the annealing program is parallel to that of process $R0$ (see Fig. 4.7). Particularly striking in Fig. 4.7 is the increase in concentration of process $R0$ defects as the free self-interstitials disappear between 33 K and 40 K. No significant number of process $R0$ defects is present either immediately after irradiation or below an annealing temperature of about 33 K (i. e., before long-range self-interstitial migration has occurred). The concentration of the $R0$ defects remains stable till approximately 60 K, and disappears completely in the 60–70 K annealing range.

The normalized relaxation strengths for the two processes $R0$ and $R4$ are, within experimental error, the same:

$$\left.\frac{\Delta_G^{<100>}}{\Delta_G^{<111>}}\right|_{R0} = 3.4 \pm 0.7 ; \quad \left.\frac{\Delta_G^{<100>}}{\Delta_G^{<111>}}\right|_{R4} = 3.4 \pm 0.4 .$$

These values indicate a defect symmetry lower than tetragonal, but not <100> orthorhombic. For process $R4$, the measured activation energy and preexponential time constant are

$$E = (0.138 \pm 0.014)\,\text{eV} \quad \text{and} \quad \tau_0 = 10^{-15\pm1}\,\text{s} .$$

The growth and recovery pattern of process $R0$ defects is again just that expected for a small interstitial cluster formed during the long-range migration and reaction of single self-interstitials. The small cluster is stable over a limited temperature range, and disappears at temperatures where thermal vibrations are sufficient to induce either its migration or disintegration.

The simultaneous recovery of processes $R0$ and $R4$ suggests that they are caused by the same defect or removed by the same defect reaction. The growth of process $R4$ could not be followed through stage I_{D+E} because of its temperature location above this stage. The first data point indicates, however, that it also grows during interstitial migration.

Processes $R0$ and $R4$ show no change during annealing in stage II_1 around 50 K where di-interstitials become mobile. Therefore the $R0$ and $R4$ defects either do not possess a large cross-section for reaction, or they are present in much greater numbers than the di-interstitials.

Process $R1$. Process $R1$ appears in the internal friction results (Fig. 4.6) near 20 K. Using the previously reported activation energy of approximately 30 meV, this process should appear at around 10 K in the elastic aftereffect results, and indeed, a small effect of the correct order of magnitude is found there.

In both techniques, process $R1$ disappears after annealing of the sample in stage I_B, where the annihilation of a close pair defect is known to occur. There are two additional relaxation processes (Fig. 4.5, at 15 K and 19 K) which disappear after annealing in stage I_B. These latter two processes do not appear in the internal friction results, because their internal friction peaks would lie above the I_B annealing stage. The existence of so many relaxation modes for a close-pair defect might be considered surprising, but it is not uncommon. Three internal friction peaks have been observed also for the I_C close pair in Ni [39].

Processes $R3$ and $R5$. Processes $R3$ and $R5$ result in very weak effects. Process $R3$ appears in the elastic aftereffect results centered about 36 K. Combining results from the elastic aftereffect and internal friction measurements, it appears to anneal in the range from 70–110 K. Its small magnitude and the proximity of the much larger effects $R2$ and $R4$ obscure its behavior during stage I_{D+E} annealing. From Fig. 4.8, we obtain for process $R3$ values of

$$E = (0.105 \pm 0.010)\,\text{eV} \quad \text{and} \quad \tau_0 = 10^{-13\pm1}\,\text{s} .$$

Process $R5$ occurs at 70 K in the internal friction measurements, and anneals somewhere between 70 K and room temperature. Its weak relaxation strength made it impossible to locate in the elastic aftereffect results. Because of the temperature range in which these two processes disappear, they are apparently due to the reorientation of small interstitial clusters. Their weak relaxation strengths indicate that the concentration of these particular sized clusters and/or their elastic anisotropies are small.

After annealing at 110 K, no anelastic processes are present in the temperature range 6 to 60 K. This observation is in agreement with diffuse x-ray scattering measurements [28], which show that by this temperature in the annealing program the interstitial clusters have collapsed to form dislocation loops.

In summary one may draw the following conclusions from the relaxation spectrum of e^- irradiated Al:

1) The elementary jump process of single self-interstitials has been identified in process $M1$ as the combined jump of the <100> split configuration. Its activation energy is,

$$E_I^R = E_I^M = 0.115\,\text{eV}$$

enabling the self-interstitial to already perform long range diffusion at low (40 K) temperatures.

2) Di-self-interstitials in the form of parallel dumbbells on nearest neighbor sites perform a two-dimensional long-range diffusion in (100) planes (process $M2$) with an activation energy slightly higher than that of the single self-interstitial,

$$E_{2I}^R = E_{2I}^M = 0.135\,\text{eV} .$$

3) The relaxation spectrum contains many relaxation processes that result from reorientations without concomitant long-range migration, i.e. localized diffusion processes. Three of them are produced by close pairs (process $R1$ plus two satellites) and five by small self-interstitial clusters.

b) Copper

A similar set of measurements was carried out on Cu samples, including a <111>-oriented sample with a Frenkel pair concentration of $c_{FD} = 5 \times 10^{-4}$ and 8.5×10^{-5} and a <100> sample with $c_{FD} = 5 \times 10^{-4}$ [26, 41]. Despite an increased sensitivity and despite a careful investigation with particular emphasis given to the temperature range of free interstitial migration between

30 and 50 K, no relaxation process — either stable or unstable with respect to annealing — was observed between 10 K and 100 K.

Previous internal friction studies of irradiated copper have reported only one point defect relaxation process, namely that due to the I_B close-pair defect [16, 40]. Because of its low activation energy of 17 meV, this process would occur in the aftereffect measurements near 5 K. Since this is below the lowest measuring temperature, it was not observed. The lack of any relaxation process between 10 and 100 K is therefore in agreement with all previous internal friction studies. Diffuse x-ray scattering [42] and the diaelastic polarizabilities (Sect. 3.1) show the existence of the <100> split configuration of the self-interstitial after low temperature electron irradiation in Cu. Therefore, a relaxation effect similar to $M1$ in Al was also expected in Cu. The failure to observe this effect may have several different explanations:

1) The magnitude of the aftereffect is less than the sensitivity of the experimental apparatus. For the <100> split interstitial, this magnitude is proportional to the square of the anisotropy, $(P_{11} - P_{22})^2$, (2.20). For the sensitivity given in the elastic aftereffect measurements the failure to detect any relaxation effect in the present experiments would be consistent with $|P_{11} - P_{22}| < 0.3$ eV. This small anisotropy is consistent with the results of the diffuse x-ray scattering experiments, which indicate a lower value for Cu than for Al, and even allow for a vanishing anisotropy within the error limits [2].

2) Mutual elastic interaction between self-interstitials, which is about five times larger in Cu than in Al [43], could reduce the relaxation strength. In this case, the defect orientations attained during migration would be primarily governed by the internal interstitial-interstitial interaction rather than by the interaction of the interstitials with the external stress field.

3) The di- and tri-interstitials could be more mobile than single interstitials. In this case the majority of single interstitials would undergo reactions with the di- and tri-interstitials before they migrate. In fact, interstitial clustering was observed to occur at much lower temperatures in Cu than in Al [2, 44]. For instance, an average size of 7 self-interstitials per cluster was found in Cu at the end of stage I at about 40 K, whereas the same size is reached in stage II at about 75 K in Al. At this level of clustering, no major relaxation effects are observable in Al either.

The self-interstitial diffusion has been observed indirectly by internal friction measurements in dilute Cu–In alloys as discussed in detail in Sect. 4.1. In this case the self-interstitials are trapped at the In atoms, and perform a localized diffusion on positions neighboring the In atom, which reveals itself by an internal friction peak. The growth of this peak in stage I_{D+E} indicates the arrival of the freely migrating self-interstitials at the In-atoms.

c) Nickel

The combined jump of self-interstitials (Fig. 2.12) in stage I_{D+E} was identified in Ni by magnetic relaxation techniques by *Forsch* et al. [45] and *Knöll* et al. [46]. The principles of these techniques are very similar to the mechanical relaxation technique except that a magnetic field is used to align the self-interstitials and that the magnetic anisotropy of a single crystal sample is used to detect this alignment. A change of the magnetic anisotropy was observed at about 55 K, whose orientation dependence is in accordance with the <100> symmetry of the dumbbell configuration. The anisotropy change coincides in temperature with a magnetic aftereffect and with the resistivity recovery stage I_{D+E}. These findings together indicate that in full analogy to process $M1$ in Al the observed processes in Ni are due to the reorientation with simultaneous migration of <100> dumbbell interstitials. The measured activation energy was found to be 0.140 eV.

d) Other fcc Metals

For fcc metals other than Al and Ni no such detailed information on the diffusional jump mechanism from relaxation studies is available. However, most fcc metals have been examined by using resistivity recovery measurements, and activation energies have been determined as given in Table 4.1. All of the metals exhibit a recovery stage I_{D+E} due to the free self-interstitial diffusion except for Au. Experiments starting at a temperature of 0.3 K did not yield any indication of stage I annealing in Au [48]. Experimental evidence indicates that the self-interstitial in Au is mobile at these low temperatures: Huang scattering measurements have determined that interstitials are in clusters after electron irradiation at 4.7 K [47].

Table 4.1. Mobility parameters for interstitial migration from [2] (T_{I_D}: temperature position of stage I_D (correlated recombination), H_I^m: activation enthalpy of interstitial migration determined in stage $I_D - I_E$)

	fcc			bcc			hcp	
Material	T_{I_D} [K]	H_I^m [eV]	Material	T_{I_D} [K]	H_I^m [eV]	Material	T_{I_D} [K]	H_I^m [eV]
Ag	28	0.088	Fe	120	0.27	Be	~30	–
Al	35	0.115	Mo	33	0.083	Cd	<4	–
Au	< 0.3	–	Nb	<8	–	Co	55	0.14
Cu	38	0.117	Ta	<8	–	Ga	33	0.073
Ni	56	0.15	V	<4	–	Gd	76	–
Pb	4	0.010	W	27	0.054	In	14	–
Pd	35	–				Mg	13	–
Pt	22	0.063				Re	83	–
Rh	32	–				Ti	120	–
						Zn	13	–
						Zr	102	0.26

4.2.2 bcc Metals

a) Iron

The resistivity recovery pattern of Fe after irradiation follows the same trend as that of the fcc metals discussed previously. Some close pair recovery substages are followed by stage I_{D+E}, the stage of free self-interstitial migration. As reviewed in [2] the dynamics and the geometry of the corresponding atomic jump processes were investigated by means of resistivity recovery studies, internal friction and magnetic relaxation measurements. The magnetic relaxation spectrum after electron irradiation exhibits a dominant magnetic aftereffect at about 110 K with a relaxation time $\tau = 35\,\mathrm{s}$. This process disappears completely upon annealing through stage I_{D+E}. The same defect causing the magnetic aftereffect was also identified by magnetic anisotropy measurements. It was found that its magnetic structure corresponds to a <110> orthorhombic symmetry in agreement with the <110> split configuration found in theoretical calculations (see Chap. 2) and x-ray scattering measurements. The activation energy for the reorientation, E^R, is reported from these relaxation experiments to be about 0.30 eV. This value agrees within the reported error, with the 0.27 eV obtained from change of slope measurements of the resistivity recovery [64]. All these experimental observations are consistent with the model developed from the calculations, that in stage I_E of Fe <110> dumbbell interstitial atoms undergo simultaneously reorientation and long-range migration by the combined jump displayed in Fig. 4.14. This is fully analogous to the combined jump of <100> dumbbells identified in Al and Ni [2].

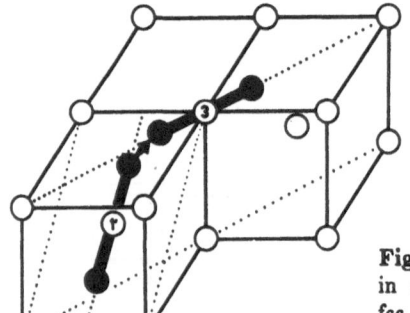

Fig. 4.14. Combined jump of a <110> dumbbell in a bcc lattice. As in the case of the <100> fcc dumbbell, simultaneously reorientation and migration occur

b) Molybdenum

The structure of the self-interstitial in Mo was identified as the <110> split configuration by Huang diffuse scattering [49]. Although its relaxation volume $(V^{\mathrm{rel}}/\Omega = 1.1)$ is quite small as compared to fcc metals, its anisotropy is quite large. In terms of Table 2.1, $P_{11} - P_{22} = 14.5\,\mathrm{eV}$ and $(P_{11} + P_{12} - 2P_{33})/2 =$

$-20.4\,\text{eV}$. The onset of self-interstitial mobility was found at 35 K from resistivity recovery [50], Huang diffuse scattering observation of self-interstitial clustering [49] and Mössbauer measurements of self-interstitial trapping at Fe^{57} atoms [51]. A migration energy of $E^M = 0.083\,\text{eV}$ is consistent with these observations. Because of the large elastic anisotropy, the self-interstitial in Mo should be an ideal object to study with the elastic aftereffect provided there is a reorientation during migration. On the basis of the anisotropy and recovery, one would expect a relaxation process with $\tau = 100\,\text{s}$ at about 30 K, with a relaxation strength $\Delta_G^{<111>} = 0.5$ for $c_{FD} = 30.0 \times 10^{-6}$, which is extraordinarily large as compared to process $M1$ in Al, for instance.

An extensive experimental search for this process was performed on <111>- and <110>-oriented samples similar to those described for Al and Cu, covering a concentration range from $c_{FD} = 12 \times 10^{-6}$ to $c_{FD} = 850 \times 10^{-6}$ [18, 52, 53, 54]. Four relaxation processes were observed in the isochronal aftereffect spectrum between 20 K and 30 K, and a similar spectrum had been observed after neutron irradiation by *Mizubayashi* et al. [55]. However, neither the elastic anisotropy, nor the annealing behavior of these relaxation lines is compatible with the properties of the self-interstitial described before. The relaxation strength is two orders of magnitude smaller than expected and the recovery occurs in two distinct stages at about 28 K and around 38 K (Fig. 4.15).

The basic conclusion from these studies must be that a reorientation of the self-interstitial does not occur during its free migration.

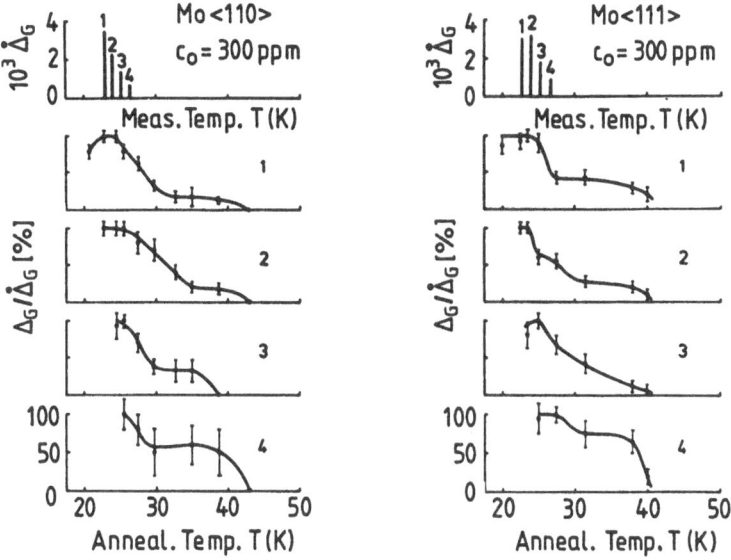

Fig. 4.15. Schematic representation of the relaxation spectrum at $\tau = 100\,\text{s}$ of electron-irradiated Mo (*top*) and of the annealing curves of the various processes (*bottom*). The left- and right-hand sides of the figure result from measurements on a <110>- and a <111>-oriented sample, respectively. For further details refer to text

Several explanations could be considered to account for this observation:

(a) Self-interstitials perform a fast pure reorientation at temperatures below 5 K, so that they are instantaneously aligned during migration. (b) The self-interstitial performs less than two jumps before the occurrence of a reaction event. (c) A strong elastic interaction between the self-interstitials prevents their alignment by an external stress field. All of these reasons can however be excluded on the basis of experiments dedicated to these questions. As to point (a), despite a low temperature location of a relaxation effect, it would be detected at higher temperatures via the $1/T$-dependent modulus effect. Such an effect was not observed. A too small number of reorientational jumps (point (b)) can be excluded on the basis of resistivity [50], *Mössbauer* [51] and diffuse x-ray scattering [42] studies, which all indicate that the self-interstitials perform more than two jumps before their reaction. Finally, concerning point (c), the elastic interaction of interstitials among each other may be drastically reduced by increasing their mutual separation as for instance by a decrease of their concentration. Measurements at self-interstitial concentrations as low as 12 atppm failed to show a relaxation effect.

In conclusion, the absence of the self-interstitial relaxation must be due to the particular nature of their jump process.

The pure translational jump of the <110> dumbbell in a <110> direction would in principle be a suitable candidate for an explanation. It leads to a one-dimensional diffusion of the self-interstitial without any reorientation. However, this jump appears highly unlikely because of the high activation energy involved. The saddle point for this jump is the octahedral position and for the α-Fe potential of Johnson its height is 1.13 eV [10]. The lowest diffusion saddle for the same potential is that for the combined jump (Fig. 4.14) with 0.21 eV. Even allowing for some differences between the true Mo potential and Johnson's Fe potential, it appears highly unlikely that the activation energy for the translational jump might be lower by a factor of 10 in Mo as compared to α-Fe.

However, in a bcc lattice there is another jump mechanism possible which is compatible with all experimental and theoretical aspects of self-interstitial migration in Mo and in particular with the relaxation measurements. For its discussion we return to a consideration of the resonant modes of the self-interstitial in the bcc structure. From Fig. 4.16 it is intuitively apparent that there are three resonant modes which may participate in the migrational process, a translational mode and two librational modes. The important point is that the two libration modes I and II are not degenerated. For the libration mode I, the atomic displacements are along a <100> lattice direction, but for the libration mode II they are along a <110> direction. In fcc metals the corresponding displacements are both along <100> directions and therefore the two modes are degenerate.

Calculations show that for α-Fe the energy necessary to excite equal vibration amplitudes is 30 % lower for libration mode II than for the libration

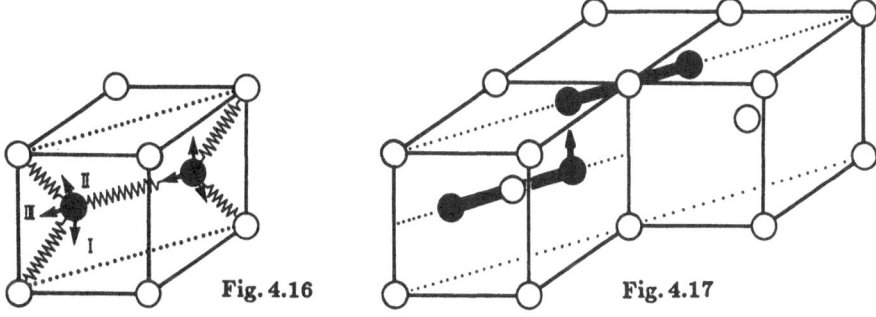

Fig. 4.16. Lattice model indicating three resonant modes of a <110> dumbbell in a bcc lattice. The two libration modes perpendicular to the dumbbell axis are not degenerated, i.e. their vibrational frequencies may be different

Fig. 4.17. "Shift" jump of a <110> dumbbell in a bcc lattice. Qualitatively, this jump can be expected to occur if the librational I mode is energetically favored over libration II, and is combined with the translation mode

mode I [56]. For this reason the dumbbell will have a strong tendency to jump *out* of its <110> habit plane and perform the combined jump shown in Fig. 4.14.

It is, however, conceivable that with slight changes of the potential the situation is reverse, namely that libration mode I is easier to excite than libration mode II. In this case the <110> dumbbell would perform particularly large vibrations *in* its <110> habit plane which in conjunction with the translation mode leads to a migrational jump in the (110) plane without rotation as shown in Fig. 4.17. We will call this jump the "shift" jump because the dumbbell axis maintains its direction during the jump. This shift jump allows the <110> dumbbell to perform a two-dimensional long-range migration without concomitant reorientation. Such a jump model would explain the absence of the mechanical relaxation effect during free interstitial migration despite the strong elastic anisotropy of the self-interstitial.

In order for this jump to actually occur two conditions have to be met. First, the migration energy for the two-dimensional diffusion, $E_{(2)}^{M}$, must be smaller than that for the three-dimensional diffusion, $E_{(3)}^{M}$. In Fe, the situation is just the opposite: from the saddle point energies one finds $E_{(2)}^{M} = 0.28\,\mathrm{eV}$ and $E_{(3)}^{M} = 0.21\,\mathrm{eV}$ [10]. Thus it follows that in Fe the three-dimensional migration is favored in accordance with the experiments. However, the difference between $E_{(2)}^{M}$ and $E_{(3)}^{M}$ is small. Therefore it is plausible to assume that for a different material, i.e. a different potential, the reversed condition could be fulfilled, $E_{(2)}^{M} < E_{(3)}^{M}$.

The second condition to be fulfilled results from the symmetry of the saddle point configuration of the two-dimensional jump, which is the <111> split configuration. Because of its trigonal symmetry, jumps into two other

71

equivalent orientations are possible. This would produce the same final configuration as attained via the reorientation jump. In order to keep the present model in accordance with the experimental observations, a strong diffusion anisotropy has to be invoked. If one assumes that eventual reorientational effects are buried in the background effects the shift jump must be favored by about 100 times over the reorientational jump.

Such a diffusion anisotropy is in principle possible. According to the diffusion theory of *Flynn* and *Stoneham* [144] the transition rate of an interstitial atom from site p into site p' is proportional to the so-called lattice overlap factor,

$$W_{pp'} \approx <a_p|a_{p'}>^2$$

where $|a_p>$ and $|a_{p'}>$ are the lattice wave functions when the interstitial atom is in position p and p', respectively.

For the shift jump, the atomic displacements for librational mode II are co-planar in a single <110> plane for both p and p', whereas they occur in two perpendicular (110) planes for the reorientational jump. Therefore it seems qualitatively clear that the overlap factor ought to be large in the first case and small in the second. Accordingly, the preexponential factor for the jump rate in the plane should be larger than that out of the plane.

A two-dimensional self-interstitial diffusion leads to special reaction schemes and in fact several "unusual" phenomena have been observed in the recovery behavior of irradiated Mo [18]. One of these is the observation of a "close pair" recovery stage above the temperature for the free migration of the self-interstitial [50]. Although other explanations may be invoked to account for this result [50] the explanation in terms of a two-dimensional self-interstitial migration is straightforward [18, 53, 54, 145]. The reaction of self-interstitials with vacancies is characterized by two reaction radii (Fig. 4.18). The first and usually the larger one is the so-called trapping radius, r_v^t. If the self-interstitials were to migrate three-dimensionally, from any position closer than r_v^t to the vacancy they would start a drift diffusion towards the vacancy with "no return". With the two-dimensional migration, however, this drift

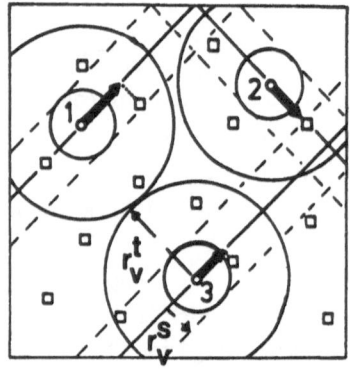

Fig. 4.18. Schematic representation of the possible reactions with vacancies of self-interstitials (*1,2, 3*) which migrate two-dimensionally in <110> planes shown in projection as the solid straight lines. r_v^s is the linear dimension characterizing the volume in which spontaneous recombination occurs. Vacancies within the range $r_v^s < r < r_v^t$ will cause the interstitials to stay trapped in a position in the plane closest to the vacancy

diffusion is suppressed for all vacancies outside the plane. The self-interstitial stays trapped in its migration plane on a position closest to the vacancy. Closer around the vacancy a volume exists defined by the second radius, r_v^s, where spontaneous recombination between the self-interstitials and vacancies occurs. If the self-interstitial diffuses two-dimensionally in one plane, it can only recombine with those vacancies which lay within a distance r_v^s of the plane but the self-interstitial may be trapped within a distance larger than r_v^s but smaller than r_v^t from the plane without recombination by a vacancy. This forms a kind of close pair. The thermally activated jump out of the plane induces the "close pair" recovery stage at a temperature slightly higher than that where the free migration of the self-interstitials occurs.

There is another quite interesting implication of two-dimensional self-interstitial diffusion. *Evans* [57, 135] has proposed that it leads to a planar ordering of voids. Voids, which are three-dimensional agglomerates of vacancies, are frequently observed in materials subjected to a high dose of high-energy particle irradiation. They result from the fact that the self-interstitials in these materials react at a somewhat higher rate with internal sinks such as dislocations than with vacancies. Since self-interstitials and vacancies are produced in equal numbers, the left over vacancies show the tendency to react with each other to coagulate into voids. Void formation is an important issue with respect to the material's properties since it results in large volume swelling. In several bcc metals such as Mo, W, Ta, Nb, and some fcc metals such as Ni and Al, a periodic arrangement of such voids may be found depending on the irradiation conditions, and several models have been put forward to account for this observation [58, 135, 146].

The basic idea of the Evans model is that the confinement of each self-interstitial to its migration plane prevents the compensation of fluctuations in the void density on adjacent planes and instead accentuates any deviation from the average void density on the plane. Because of the six equivalent (110) crystal planes, a body centered cubic arrangement of void lattice sites emerges.

The effect has been demonstrated by Monte Carlo calculation. An example is shown in Fig. 4.19, taken from [135]. 125 voids of radius 0.7 nm were

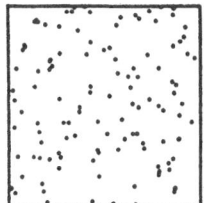 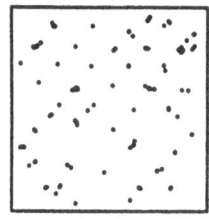 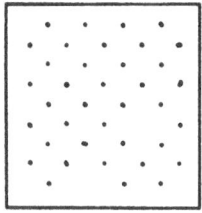

Fig. 4.19. Monte Carlo simulation of the formation of a void lattice in a bcc crystal, in which the self-interstitials are assumed to migrate two-dimensionally. Shown is a projection into a crystallographic <100> direction. One recognizes the development from a spatially random void distribution into a bcc void lattice with increasing irradiation dose. From [135]

placed randomly in a model crystal of edge length 31.4 nm. The positions of the void were followed with increasing irradiation dose. Figure 4.19 shows a projection along a <001> direction at the beginning and at two irradiation doses. The voids are marked by black dots. The formation of a void lattice with increasing dose is evident. Projections along different directions show that the void lattice has the bcc structure.

c) Other bcc Metals

As in Mo, two-dimensional self-interstitial diffusion may possibly also occur in W. This conclusion is suggested by the similarities of W and Mo in their resistivity recoveries after electron irradiation [60] and their mechanical relaxation spectra after neutron irradiation [59].

The V-metals group (Nb, Ta, V) shows self-interstitial migration at temperatures below 8 K [60]. Such low temperatures indicate low migration energies. Details of the migrational jumps in these metals have not been determined.

4.2.3 hcp Metals

Magnetic aftereffect measurements on Co [147–149] and internal friction measurements on Zr and Ti [150, 151] have been reported and show relaxation effects from self-interstitials in these metals. However, definite assignments to specific configurations and specific jump geometries have not been made.

4.2.4 Data Collection

In addition to the use of the relaxation techniques, the diffusion of self-interstitials has been investigated in many metals by indirect methods. Among these are the change of slope technique as applied to resistivity annealing, and the onset of self-interstitial clustering as observed by diffuse x-ray scattering. Such measurements provide figures for the activation parameters of the diffusion process or — at least — the temperature location of annealing stage I_{D+E} where the free migration of self-interstitials occurs.

Data for the temperature location of stage I_D, T_{I_D}, and for the migration enthalpies of the self-interstitial, E_I^m, are collected in Table 4.1 as taken from [2]. Similar compilations may be found in [30] and [162].

The T_{I_D}-data may be taken for a rough estimate of E_I^m in those cases where E_I^m has not been determined directly. Since T_{I_D} is found to be concentration-independent, its value is directly related to the corresponding activation enthalpy by $E_I^m = k \cdot T_{I_D}$ [61, 62]. For those metals where both, T_{I_D} and E_I^m, are reported, k can be evaluated as $(3.0 \pm 0.3) \times 10^{-3}$ eV/K for the fcc metals and $(2.25 \pm 0.25) \times 10^{-3}$ eV/K for the bcc metals.

Although T_{I_D} has a rather wide range, from 0.3 K in Au to 120 K in Fe and Ti, self-interstitial diffusion generally occurs at "low" temperatures as compared to vacancy diffusion which occurs typically above 200 K for most of the cubic metals [2]. Compared to self-diffusion processes, which are predominantly carried by vacancy fluxes, diffusion via self-interstitials would occur at a much faster rate at the same temperature. This has important implications for atomic transport in irradiated materials, and it may have consequences with respect to self-diffusion at high temperatures [23]. The relevant quantities to consider for a comparison of the self-interstitial and vacancy contribution to self-diffusion are the values of $(E^f + E^m)$ for the self-interstitials and vacancies and the self-diffusion energy Q^{SD}, where E^f and E^m are the formation and migration energies, respectively. In the "classical" picture of self-diffusion, Q^{SD} is identified with $(E_V^f + E_V^m)$ where E_V^f and E_V^m are the formation and migration energy of the vacancy, respectively. The argument behind this is that the formation energy of the vacancy is much smaller than the formation energy of the self-interstitial E_I^f, so that the concentration of self-interstitials in thermal equilibrium is always negligibly small. Since, however, the migration energy of self-interstitials, E_I^m, is much smaller than the vacancy migration energy, this argument may not hold for all metals. Thus it has been speculated that self-interstitials possibly contribute to self-diffusion at high temperatures in Cu and Pt [23] and in the refractory bcc metals like Mo and W [63]. A reliable estimate of this contribution is difficult to make since the error limits of the formation energies of the self-interstitials in metals are rather large.

5. Experimental Results for Dilute Alloys

In the following sections the experimental findings for interstitial atoms in dilute alloys are compared to the results of the simple modelling described in Chap. 2.7. In addition, values for binding energies and trapping radii are collected and the conditions for their applicability in model calculations are discussed.

Since the only general model available for the interaction between a self-interstitial and a solute atom is based upon atomic size differences, one can try to correlate the measured binding energies and trapping radii with the impurity size defined as

$$\delta_{SA} = \frac{1}{\Omega_0} \cdot \frac{\delta\Omega}{\delta c_t} \tag{5.1}$$

where Ω_0 is the atomic volume of the solvent, and $\delta\Omega/\delta c_t$ is the change of the atomic volume, Ω, with the concentration, c_t, of the solute atoms. δ_{SA} values have been collected by *King* [75].

5.1 Copper Alloys

Table 5.1 summarizes the results obtained for Cu alloys. The trapping radii, r_t, are given in terms of r_v, the recombination radius for spontaneous Frenkel pair annihilation [76]. Three types or groups of trapping behavior can be distinguished:

Group I. The solute atoms in this alloy group are oversized. All of these solute atoms trap self-interstitial atoms up to temperatures above 50 K. In all alloys of group I final detrapping occurs in recovery stage *II*. The binding energies fall in a narrow range from 0.2 to 0.3 eV. There is no obvious correlation of E^b or of r_t/r_v with the atomic size differences.

Group II. Be is the only known representative of undersized solute atoms that trap self-interstitials in Cu. The complexes are stable up to recovery stage *III* ($\sim$200 K).

Group III. This group contains both under- and oversized solute atoms which show little or no self-interstitial trapping above 50 K. Consequently, the possible upper limits for binding energies and trapping radii are very small.

Table 5.1. Trapping properties of Cu alloys: δ, size factor: E^b, binding energies of the most stable SI complex: r_t/r_v trapping radii in units of the recombination radius, r_v

Group	solid atoms	δ [%]	E^b [eV]	r_t/r_v
I) Trapping observed $\delta > 0$	Si	5.08	0.34	0.6
	Cr	19.72	0.24	
	Ti	25.74	0.28	1.3
	Pd	27.96	0.18	0.5
	Mn	34.19	0.20	1.3
	Ag	43.52	0.28	1.3
	Au	47.59	0.21	0.8
	Mg	50.80	0.31	
	In	79.03	0.32	1.5
	Sb	91.87	0.20	1.3
II) Trapping observed $\delta < 0$	Be	-26.45	>0.50	1.4
III) No trapping observed	Ni	-8.45	<0.03	0
	Co	-3.78	<0.03	0
	Fe	$+4.57$	<0.10	
	Zn	$+17.10$	<0.08	0.07
	Ge	$+27.77$	<0.08	0.09

It is obvious from Table 5.1 that many aspects of the simple model based upon size difference are not verified by the experiments. Si and Fe, Cr and Zn and Pd and Ge, which are similarly oversized, behave entirely differently. Ni and Co are expected to bind the self-interstitial atoms very tightly since they are undersized, but they do not. Finally, as mentioned before, no correlation of E^b and r_t with δ_{SA} is observed. Nevertheless, there are many common features among group I self-interstitial atoms and characteristic differences between these and Cu–Be which are worth mentioning and which will help to elucidate characteristic aspects of self-interstitial solute interactions in fcc alloys.

Group I Alloys. A representative example of the trapping behavior of group I self-interstitial atoms is Cu–In. This alloy has been studied using four different experimental techniques: resistivity recovery and damage rate [78], internal friction [80], and perturbed angular correlation [81]. Results of these measurements are shown in Figs. 5.1–3. Resistivity recovery due to reactions of SI complexes occurs in two pronounced stages at about 70 K and 150 K, which according to the notation used by *Cannon* and *Sosin* [82] are denoted by II_B and II_C, respectively (Fig. 5.1). The latter, II_C, is the stage of central interest in the present context. It occurs in a similar manner in all group I alloys [71, 77–88]. In Cu–Ag and Cu–Au it has been identified by Cannon

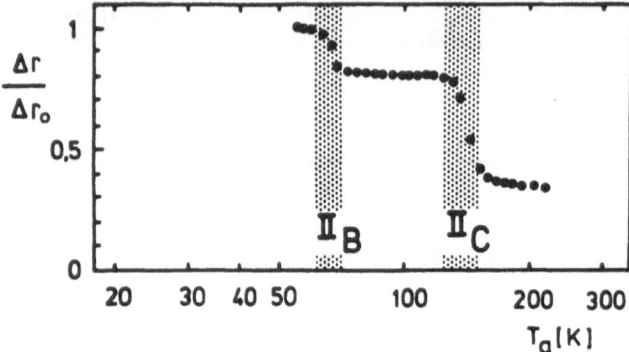

Fig. 5.1. Recovery of the residual electrical resistivity change of CuIn, Δr, normalized to the total radiation-induced resistivity change, Δr_0, during isochronous annealing after electron irradiation. T_a is the annealing temperature, II_B and II_C are the stage assignments following Cannon and Sosin

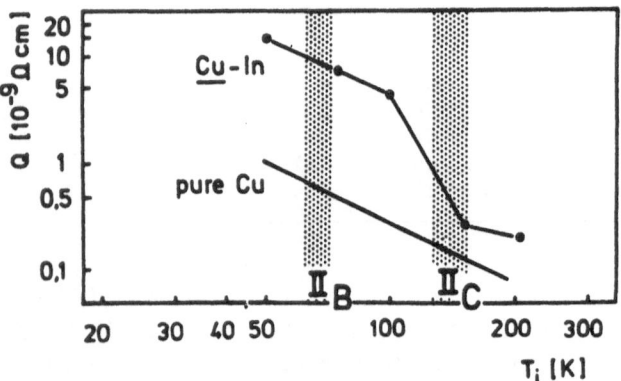

Fig. 5.2. Temperature dependence of the trapping rate $K/(4\pi D_c r_v/\varrho_t) = Q$ of Cu and CuIn during electron irradiation at temperatures T_i

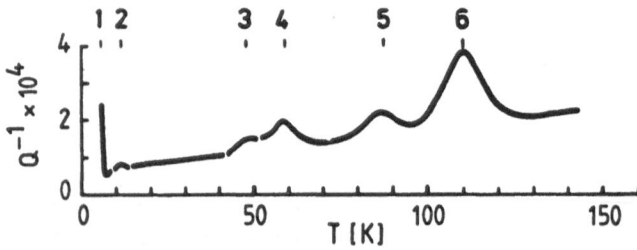

Fig. 5.3. Internal friction spectrum of CuIn (400 atppm) after 3 MeV electron irradiation. The peaks are plotted with their greatest heights obtained during the isochronous annealing treatment (cf. Fig. 5.4). Measurements were taken in a vibrating reed device on a polycrystalline sample, i. e. C and C' mode responses may both contribute to the spectrum

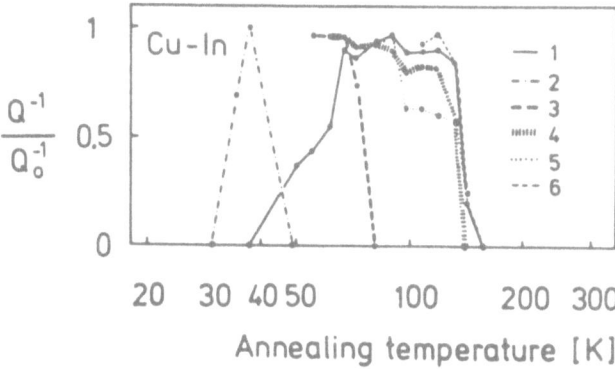

Fig. 5.4. Annealing behavior of the CuIn internal friction peaks of Fig. 5.3

and Sosin as the dissociation of SI complexes. By analogy the II_C stages of the other group I alloys are interpreted similarly. For Cu–In, this interpretation is corroborated by the two other studies, the damage rate and internal friction experiments. The trapping rate K in units of $4\pi D_0 r_v/\varrho_t$ as determined from damage rate measurements shows a step-like drop as a function of irradiation temperature (Fig. 5.2). This indicates the strong trapping capability of the In atoms for the migrating self-interstitial atom at temperatures below II_C. At temperatures above stage II_C, the trapping term essentially decreases to the same values found for pure Cu. This demonstrates directly that at temperatures above stage II_C the In atoms lose their trapping capability. Additional details about the SI complexes in Cu–In have been obtained from internal friction studies. In the internal-friction spectrum of electron-irradiated Cu–In, Fig. 5.3, six distinct maxima occur. The normalized heights of these are shown in Fig. 5.4 as a function of the annealing temperature, T_a. Peaks 1, 4, 5, 6 disappear in the temperature regime around II_C. The defects responsible for these maxima do not anneal exactly at the same temperature but slightly separated from each other, as indicated in Fig. 5.4. *Kollers* et al. [80] concluded that not one but at least three different kinds of SI complexes disappear in the temperature regime around II_C. Above II_C temperature, no further internal friction peaks are observed. This result is consistent with the other observations that final detrapping has occurred in II_C. The defects annealing in II_C do not give rise to any defect lines in the perturbed angular correlation spectrum (Fig. 5.5) and remain invisible in these measurements. *Canon* and *Sosin* [82] have analysed stage II_C of Cu–Ag and Cu–Au in great detail and find:

i) The processes inherent in II_C obey first-order kinetics.

ii) The average number of jumps, N, the defects have performed in order to cause the observed resistivity recovery, amounts to $N \simeq 1$. This number is characteristic of a mechanism where after one dissociation jump the liberated self-interstitial diffuses practically instantaneously towards another more sta-

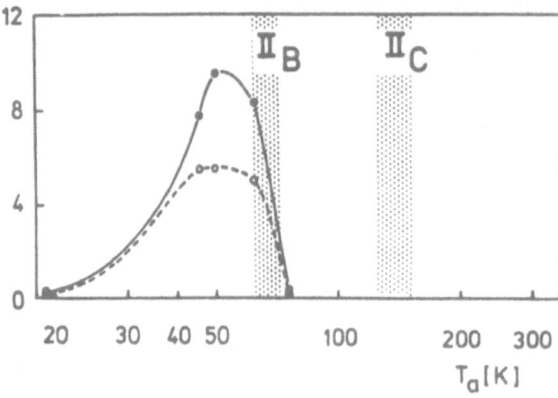

Fig. 5.5. Intensity of defect lines in arbitrary units observed in the perturbed angular correlation spectrum of electron-irradiated CuIn [81]. A line correlated with stage II_C recovery was not identified

ble sink. The relation between the dissociation energy E^d and the binding energy E^b of the SI complexes is given by

$$E^d = E^b + E_I^m$$

where E_I^m is the migration energy of the self-interstitials. The binding energies E^b may be estimated from the peak temperature of stage II_C, T_{II_C}, using the relation [77]

$$E^b = 3 \times 10^{-3} (\text{eV/K}) \, T_{II_C} - E_I^m . \qquad (5.2)$$

The values of E^b given in Table 5.1 do not point to any systematic relationship between E^b and the atomic size δ_{SA}. For instance, the slightly oversized Si atom exhibits the same binding energy as the greatly oversized In atom.

The recovery stage II_C is not the only recovery stage found for group I atoms. At least one more pronounced stage, II_B, is observed. The perturbed angular correlation investigations show that two different defect types anneal simultaneously in stage II_B of CuIn (Fig. 5.5). In the internal friction measurements of CuIn relaxation process 1 is observed to grow partly in stage II_B. Although both experiments show pronounced defect reactions in stage II_B, a conclusive interpretation has not yet been given about the nature of II_B.

The internal friction measurements indicate that in addition to the defect reactions occurring in resistivity recovery stage II_B and II_C several other defects come and go throughout the whole regime of stage II. An SI complex causing relaxation process 2 is created in stage I_D-I_E ($\sim$36 K) and disappears at about 43 K: peak 1 already grows markedly at temperatures below II_B and peak 3 disappears just above II_B at about 80 K. Some of the complexes which disappear finally in stage II_C at about 150 K, show a gradual and partial recovery between 80 K and 150 K.

The trapping radii r_t/r_v obtained from damage rate measurements obey an apparent T^{-2} dependence (Fig. 5.2) in Cu–In and all other group I alloys, except Si. This T^{-2} dependence has been explained by *Wollenberger* [89] by the assumption that many different SI complexes produce a quasi-continuous spectrum of binding energies. Therefore a decreasing number of SI complexes remain in the sample with increasing irradiation temperature.

All these experimental observations point to the fact that an ensemble of differently structured SI complexes may be created during irradiation, even if the self-interstitial atom concentration, c_I, is smaller than the solute atom concentration, c_{SA}. The existence of such a complex ensemble may result from two different sources: (i) complexes of different size may be created by multiple trapping. As an example, *Kollers* [93] has shown that the internal friction spectrum of Cu–In with $c_I/c_{SA} = 0.16$ can be accounted for by reorientations of complexes containing from one to four self-interstitial atoms bound to one solute atom. The concentration of SI complexes employed to give this interpretation were obtained from a computer simulation where the production and the reactions of the self-interstitials, vacancies and the various complexes were described by an appropriate set of rate equations as discussed by *Bewerunge* [71] and *Becker* et al. [72]. (ii) The single SI complex (and possibly the larger complexes, too) may be present in several configurations with different structures and stabilities. Therefore the binding energies E^b derived from Table 5.1 belong to those complexes with the highest binding energies.

Group II Cu–Be. The behavior of Cu–Be is very different from that of group I solute atoms: trapping of self-interstitial atoms at solute atoms is observed up to stage *III*. Although partial recovery and configurational changes of SI complexes occur in stage *II* [85, 89, 90], a substantial fraction of SI complexes remains until about 250 K, where they are annihilated by migrating vacancies. From the arguments given before, the binding energy for these stable complexes must exceed a value of 0.5 eV or their migration energy must be larger than 0.6 eV.

Bartels et al. [91] found from damage rate measurements that the Be concentration in solution, c_{SA}, decreased linearly with the irradiation dose and from the temperature dependence of the "removal rates", dc_{SA}/dF, where F is the irradiation dose, they concluded an upper limit of $E^m \leq 0.6$ eV for the migration enthalphy of the complex. They propose an interstitial transport of Be atoms in order to explain the observed Be depletion under irradiation.

Group III Alloys. No additional resistivity recovery stages were observed in group III alloys in stage *II*, and the trapping rates observed at 50 K were practically the same as those in pure Cu. In principle it could be argued that SI complexes are formed but are as mobile as the self-interstitial atoms so that they escape from experimental observation. Nevertheless they would cause segregation. Although there is no evidence to exclude this assumption, there is also no indication in support of it.

5.2 Aluminum Alloys

The same grouping as used for the Cu alloys is also used for Al alloys. The result is shown in Table 5.2.

Table 5.2. Trapping properties of Al alloys: δ, size factor: E^b, binding energies of the most stable SI complex: r_t/r_v, trapping radii in units of the recombination radius, r_v

Group	solid atoms	δ [%]	E^b [eV]	r_t/r_v
I) Trapping	Mg	40.82	0.37	1.5
observed	Ge	13.13	≥ 0.50	2.0
$\delta > 0$	Ga	4.94	≥ 0.50	
	Ag	0.12	≥ 0.50	3.0
II) Trapping	Cr	-57.23	≥ 0.50	0.3
observed	Mn	-46.81	≥ 0.50	
$\delta < 0$	V	-41.42	≥ 0.50	1.0
	Cu	-37.77	≥ 0.50	
	Co	-39.95	≥ 0.50	
	Fe	-29.09	≥ 0.50	
	Ti	-15.06	0.22	1.5
	Zn	-5.74	≥ 0.50	3.0

Group I Alloys. Group I includes those alloys which contain oversized solute atoms and show trapping of the self-interstitial atoms. It is obvious from Table 5.2 that these alloys behave differently from the alloys of the corresponding copper group. Resistivity recovery occurs in Al–Mg in three stages at 80 K, 140 K and 160 K, Fig. 5.6 [80, 92, 94]. At temperatures above 160 K, the resistivity recovery curve of Al–Mg resembles within experimental error the recovery curve of pure Al. This indicates that in this latter recovery stage final detrapping has occurred. The same conclusion follows from the temperature dependence of the trapping term $Q = c_t r_t/c_v$, which shows a decrease by an order of magnitude between 100 K and 180 K to the same value as observed in pure Al [77]. The internal friction spectrum (Figs. 5.7 and 5.8) consists of five distinct peaks due to SI complexes which disappear in two groups. Peaks 2, 3, 4 vanish after annealing at 140 K and the remaining two peaks 1 and 5 disappear at 160 K. Resistivity recovery stages within the stage II regime also occur for most of the other group I alloys [92] but there is no substage due to final detrapping as in Al–Mg. For example, damage rates, perturbed angular correlation and channelling measurements on aluminum containing Ag [92, 95, 96], Ge [92, 97], In [98], and Ga [96] demonstrate that there are SI complexes which remain stable and immobile until stage III, at about 200 K. Therefore, as shown in Table 5.2, only lower limits for either E^b or E^m can be given for the most stable complexes.

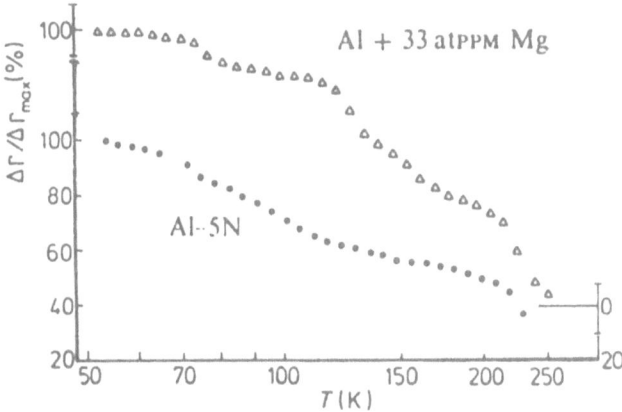

Fig. 5.6. Resistivity recovery in pure Al and an AlMg (33 atppm) alloy after 3 MeV electron irradiation. Data taken from [92]

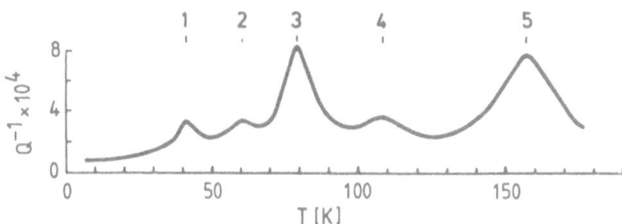

Fig. 5.7. Internal friction spectrum of an AlMg (250 atppm) alloy after 3 MeV electron irradiation and annealing at 130 K. Measurements were taken by means of the torsion pendulum of Fig. 3.5, the sample was polycrystalline material, i.e. both, C and C' type relaxations may contribute. The annealing behavior of the five peaks is displayed in Fig. 5.8

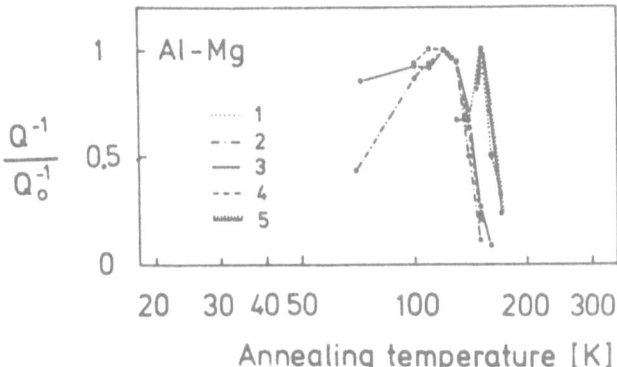

Fig. 5.8. The change of the normalized peak heights with annealing temperature of the five internal friction peaks observed in electron irradiated AlMg (250 atppm), cf. Fig. 5.7

The magnitude of the trapping radii r_t/r_v determined from the damage rate studies cannot be explained by elastic dipole interaction between self-interstitial atom and solute atom (2.31) and (2.32). Since Al is almost elastically isotropic in contrast to the very anisotropic Cu, the trapping radii should typically be about a factor of two smaller in Al alloys compared to Cu alloys containing equally under- or oversized solute atoms. An extreme case is Ag in Al: although Ag has a very small size difference from Al, it possesses the largest trapping radius of group I and a large binding energy. Furthermore, *Dworschak* et al. [79] have pointed out that the trapping radii decrease with increasing size difference instead of increasing as required by (2.31). Apparently, however, the evaluation of trapping radii in Al alloys from resistivity measurements is not without difficulty. This is mainly due to an overlap of stages I_C (close pairs) and I_D (correlated recovery) and due to a possible change of the Frenkel pair resistivity during trapping as discussed by *Maury* and *Wollenberger* et al. [139, 140].

Group II Alloys. The solute atoms listed in this group are all undersized with respect to the Al atom and, except Ti, do not show final detrapping in stage II. The lower limit for the binding enthalpies or migration enthalpies for the most stable complexes can therefore be estimated as about 0.5 eV. In Al–Ti, detrapping occurs at about 110 K, but it has not yet been shown whether the observed resistivity recovery is due to a release process or to a migration of the SI complex.

The discussion of the trapping radii given for the group I alloys is also valid for group II: the trapping radii r_t are too large for elastic dipole interaction and decrease with increasing size difference [79].

Direct experimental evidence for the cage motion has been found in Al–Fe by both internal friction and Mössbauer measurements [99–101]. The internal friction peak 1 at 8 K indicates a local reorientation of the SI complex with an activation energy of about 13 meV (Fig. 5.9). Internal friction peak 4 at about 50 K parallels the annealing behavior of the first (Fig. 5.10). The two peaks were interpreted as a peak doublet resulting from two different jump modes of one SI complex [66]. This so-called "frozen free split" phenomenon

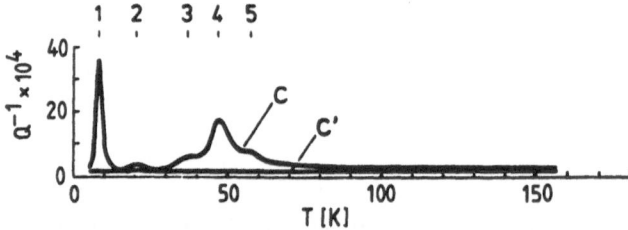

Fig. 5.9. Internal friction spectrum of AlFe (250 atppm) after electron irradiation and annealing at about 50 K. By employing different single crystalline and polycrystalline samples in the torsion pendulum of Fig. 3.5 the responses to the C and C' mode deformation were obtained separately as shown. The initial Frenkel pair concentration was 400 atppm

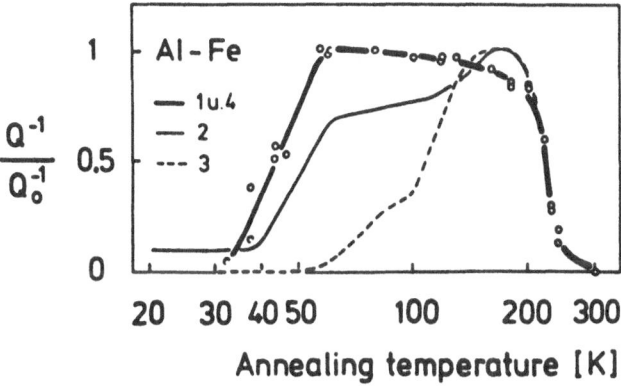

Fig. 5.10. Change of the normalized relaxation strengths with annealing temperature of the AlFe relaxations shown in the previous Fig. 5.9

was explained by the cage shown in Fig. 2.19b. According to this explanation the 8 K peak results from jumps of the Fe atom within each of the triangles while the 50 K peak occurs by jumps between different triangles.

In the Mössbauer experiments, a step-like drop of the Debye-Waller factor at about 16 K was attributed to the same defect and the jump process which causes the 8 K internal-friction peak. However, no evidence for a second jump mode of this defect was found in the Mössbauer experiment. Because of this and because of the observed orientation dependence of the Debye-Waller factor a cage as shown in Fig. 2.19c was employed to explain the Mössbauer results [100–103].

Another contradiction exists between the models used to interpret the internal friction and the Mössbauer results and the <100> mixed dumbbell model used to interpret the channelling data on Al–Fe [96]. Mainly for historical reasons, the other models have not been tested and cannot be excluded. Because of the much higher defect concentrations used in the channelling experiments, different defect ensembles may have been observed in the channelling samples than in the internal friction and Mössbauer samples.

Ultrasonic attenuation studies of electron-irradiated Al–Fe alloys [141] showed, in addition to peaks 1–5 known from the internal friction work, two further peaks 1a and 1c. Peak 1c exhibits a <100> tetragonal symmetry as for instance compatible with the <100> mixed dumbbell. On the other hand, peak 1b, which corresponds to internal friction peak 1 and peak 4 with their trigonal or near trigonal symmetry are also present. In total an even greater complexity of the defect pattern than before has emerged from the ultrasonic studies. Therefore the conclusion concerning Al–Fe is as expressed in the discussion of *Setser* et al. [141] that "in spite of the fact that the system has been studied using many different techniques ..., the identification of the principal defects remains in doubt".

Indications of the cage motions of SI complexes at low temperatures have also been detected in Al–Mn and Al–Zn by ultrasonic studies after electron irradiation [104–107].

The principle peak at 4 K in Al–Zn disappears at about 130 K. In order to distinguish between a possible complex migration or dissociation mechanism underlying this reaction, *Wallace* et al. [142] added small amounts of Fe to the Al–Zn alloy. If at 130 K a complex-dissociation were to occur, the Al–Fe peaks discussed before should grow at this temperature. The expected growth was not observed and therefore a dissociation mechanism operative for the reaction at 130 K was excluded.

In conclusion the following can be said about the interaction between self-interstitials and solutes in Al alloys:

i)   The binding energies with oversized solute atoms are larger than those of comparably oversized solute atoms in copper and larger than expected for elastic interaction. With the exception of Al–Mg and Al–Ti they exceed 0.5 eV. Since the actual values cannot be determined, nothing can be said about their correlation with size difference.

ii) The magnitude and trend of the trapping radii with size difference excludes elastic interaction for the long-range interaction between self-interstitial and solute atom.

iii) The formation of an ensemble of differently structured SI complexes occurs at low self-interstitial and solute atom concentrations.

5.3   Nickel Alloys

Ni-based alloys are of particular interest in the present context, since this system shows a marked influence of solute atom on radiation-induced segregation and void swelling [108]. Basic studies of SI complexes in nickel alloys have been performed by magnetic aftereffect and internal friction measurements in Ni–Mn, Ni–Si, Ni–C and Ni–Mo after electron irradiation [109] and resistivity studies on Ni containing either Co, Fe, Cu or Si [110] and diffuse x-ray scattering studies in NiSi and NiGe [111, 112].

The formation of different SI complexes with different stabilities was observed in NiSi. After self-interstitial migration at the end of stage *I*, the build-up of a magnetic aftereffect was observed, which is absent in pure Ni [109]. It was attributed to the magnetic-field-induced reorientation of self-interstitials trapped at Si atoms. The defect is not very stable and disappears upon annealing at about 77 K. Three additional defect configurations were found in the internal friction spectrum of NiSi [109], which were attributed to increasingly larger SI complexes. These complexes anneal at 120 K, 150 K and 200 K. In a different but similar internal friction experiment, only two peaks, annealing at 120 K and 200 K, could be detected (Figs. 5.11 and 5.12) [113]. From the diffuse x-ray work [111, 112], it was concluded that the simple

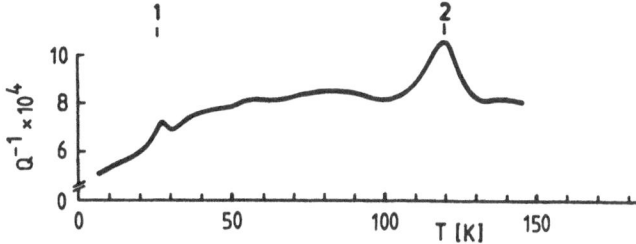

Fig. 5.11. Internal friction spectrum of polycrystalline NiSi (100 atppm) after low temperature electron irradiation ($c_{FD} = 300$ atppm) and annealing at 100 K

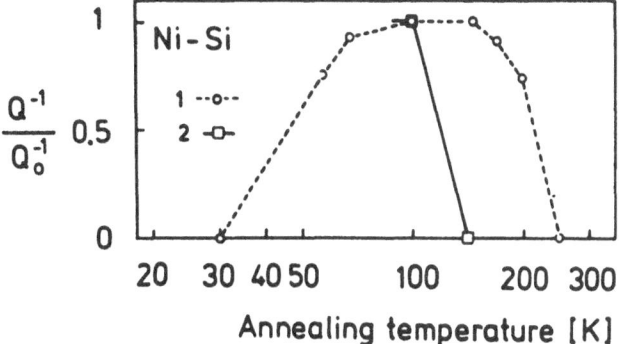

Fig. 5.12. Change of the normalized relaxation strengths with annealing temperature of the NiSi peaks shown in Fig. 5.11

complex consisting of one self-interstitial and one solute atom cannot be stable in NiSi and NiGe beyond stage I, i.e. about $T = 65$ K. Only clusters are observed which are practically stable up to the final recovery stage slightly above 200 K.

A different interpretation of the trapping behavior of NiSi has been given by *Bartels* et al. on the basis of resistivity recovery and damage rate studies [143]. They concluded that the trapped single interstitials are immobile and stable up to 105 K, where the SI complex migrates. This interpretation is clearly at variance with the interpretation of the diffuse x-ray scattering (trapped single self-interstitial unstable or mobile for $T \geq 65$ K) and that of the magnetic aftereffect studies (trapped single self-interstitial migrates or dissociates at 77 K).

Irrespective of the precise temperature location of this process, in an experiment similar to that performed on Cu–Be [91], *Weigert* [114] found a bulk depletion of Ni–Si after electron irradiation, which was attributed to a fast interstitial transport of Si atoms for instance as mixed dumbbells.

The defect pattern in Ni–Mn and Ni–Mo, which are both similarly oversized, looks much simpler than that of Ni–Si. In both alloys only one pronounced internal friction peak occurs, which in the case of Ni–Mn is attributed to the same defect that produces a magnetic aftereffect and disappears upon annealing at 130 K. The binding energies in the other Ni alloys are apparently very small: values of 0.03 eV for Co, 0.04 eV for Fe and Sb and 0.05 eV for Cu are reported by *Oddou* [110].

5.4   Au- and Pt-Based Alloys

Experimental evidence for trapping of self-interstitials at solute atoms was found in a number of Au alloys [116–118] and Pt–Au [119–125]. In the latter case, an analysis similar to that of Cannon and Sosin was performed on the resistivity recovery stages II_B and II_C after electron irradiation. II_C was attributed to final detrapping, by analogy to group I Cu alloys, and a binding energy of $E^b = 0.21$ eV for the most stable complex was deduced. These results were corroborated by the field ion microscopy results of *Wei* et al. [124, 125], who observed the arrival of self-interstitials at the specimen surface at both stage II_B and stage II_C temperatures.

5.5   Ag Alloys

Much recent experimental work has been focused on Ag alloys which potentially offer many advantages over alloys based on Al and Cu. With respect to the size effect a whole set of alloys with different size factors ranging from $\delta_{SA} = -28\%$ for Cu to $\delta_{SA} = +71\%$ for Bi can be realized. Many of these systems possess a sufficiently high solubility so that the preparation of a homogeneous solid solution does not involve any great problems. Resistivity measurements have been performed on Ag–Al, Au, Cu, Ni, Pt, Zn after electron irradiation [152–155] and on Ag–Cu, Co, Au, In, Mg after neutron irradiation [156]. Internal friction studies have been carried out on electron irradiated Ag–Al, Au, Cu, Pt, Zn [115] and neutron irradiated Ag–Cu, Ge, Ni, Pb, Sb, Su, Zn [157]. The size factors of Mn, Cu, Pt, Zn, Al, Au atoms in the Ag alloys employed in internal friction experiments after electron irradiation are $\delta_{SA}/\% = -28, -20, -14, -9, -2$, respectively.

Figures 5.13 and 5.14 show the recovery of the electrical resistivity and of the diaelastic polarization for pure Ag and the five dilute Ag alloys, respectively. Trapping of interstitials is indicated by a retardation of stage I recovery. A pronounced retardation of the resistivity recovery occurs for Ag with Al, Cu, Zn, but only small effects are observed in AgPt and AgAu. These results of the annealing of the electrical resistivity are in agreement with earlier investigations after electron [152, 153] and neutron [156] irradiation. Trapping in the two latter alloys is more clearly revealed by the pronounced retardation of the recovery of the diaelastic polarization. Major recovery occurs in Ag–Al, Cu, Zn in substages at about 100 K and 170 K, see Fig. 5.14.

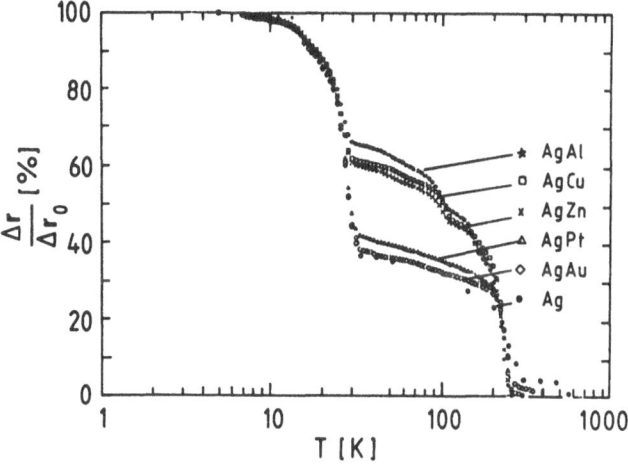

Fig. 5.13. Recovery of the electrical resistivity of Ag and Ag alloys after electron irradiation. Isochronous holding times were 10 min. From [115]

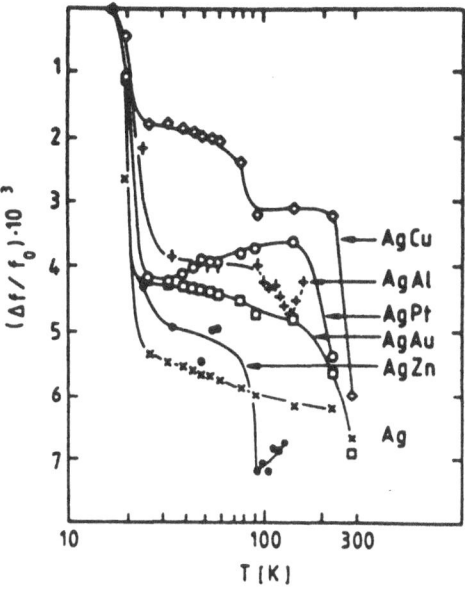

Fig. 5.14. Recovery of the diaelastic polarization as measured by the relative change $(f - f_0)/f_0$ of the vibrational frequency of the vibrating reed samples. The reference, f_0, is the frequency immediately after the irradiation. Effective isochronous holding times were about 30 min

The internal friction spectrum of AgCu exhibits two peaks located at 25 K and 40 K for a measuring frequency, f, of 150 Hz (Fig. 5.15). The build-up of the 25 K peak occurs in stage I_{D+E} at about 25 K, the build-up of the 40 K peak occurs at a temperature below 40 K. The two peaks show identical recovery behavior decaying in a parallel fashion in a single stage at 48 K (Fig. 5.16).

No internal friction peaks specific for the other solutes Al, Au, Pt, Zn were observed in the spectra of the respective alloys [115].

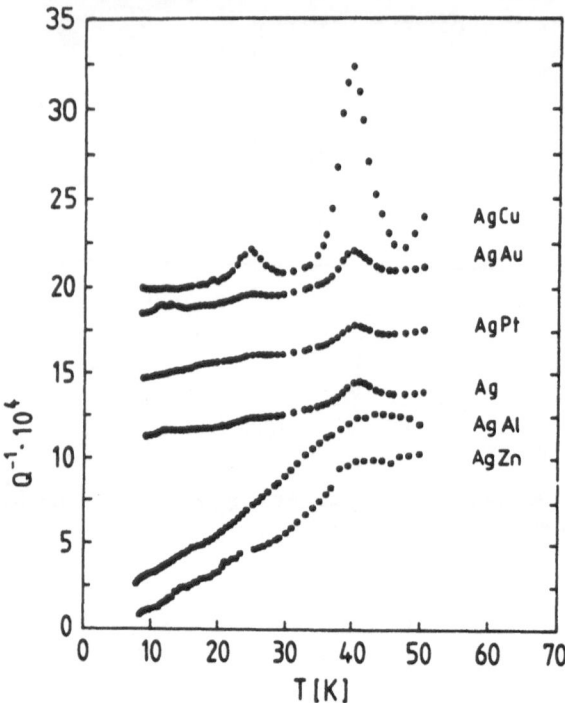

Fig. 5.15. Internal friction spectra of Ag and Ag alloys as indicated, after 3 MeV electron irradiation and annealing at 40 K. The Frenkel pair concentration was $c_{FD} = 250$ atppm, the solute concentrations within $\pm 10\%$ at the nominal concentration of $c_{SA} = 1000$ atppm. The small peaks at 25 K and 40 K in the samples other than AgCu are due to an inadvertent, small Cu contamination of these materials

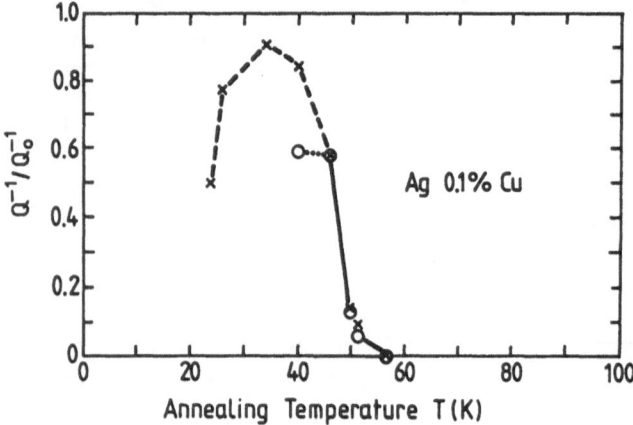

Fig. 5.16. Annealing behavior of the 25 K (∗) and the 40 K (○) internal peaks in AgCu shown in Fig. 5.15

The parallel recovery of the two internal friction peaks in AgCu suggests that they should be attributed to the motion of an off-center Cu atom in a two-mode cage as for instance discussed for AlFe (Fig. 2.19b). In this model, the peak at 25 K would be caused by a rearrangement of the site occupations by jumps within a triangle only, whereas the 40 K peak results from an additional rearrangement of the site occupations facilitated by jumps between the different triangles (Fig. 2.19b). The disappearance of the peak at 48 K occurs without any significant annealing of the electrical resistivity or diaelastic polarization. Therefore a configurational change is likely to be responsible for this reaction, for instance a transition of the Cu atom into the center of the octahedral cell. As an alternative explanation it could be assumed that this complex migrates and reacts predominantly with further Cu atoms, rather than vacancies, since $c_v \ll c_{Cu}$ at this stage [159].

The SI complexes formed in the other dilute Ag alloys do not give rise to internal friction peaks, which either means that for the most common complex configurations the deviation from cubic symmetry is too small to be detected by internal friction, or that the related activation is too high.

In contrast to this, internal friction peaks were found in all Ag alloys after neutron irradiation [157]. This result is not surprising, since it can be accounted for by a more complex defect pattern induced by the fast neutrons. One might speculate that very small solutes, i.e. $\delta_{SA} < -0.25$, are the most favorable candidates for the formation of "cages" comprised of highly symmetrical off-center sites, as also the examples of Al–Fe and Al–Mn demonstrate. On the other hand, the slightly undersized solute Zn in Al shows well pronounced relaxation effects. Therefore the results of anelastic measurements on the various Al, Cu and Ag alloys lead to the conclusion that each system has its own characteristic relaxation spectrum with little recognizable resemblance to the others and little relationship to any simple physical solute specific parameter or interaction potential.

5.6   FeNi

FeNi is probably the most widely investigated example of a bcc alloy. It has been subjected to electrical resistivity [115, 163, 164], magnetic aftereffect [165, 166] and internal friction [115, 159, 167] studies.

Figure 5.17 shows internal friction peaks (left part) and their behavior upon isochronous annealing (right part) as observed in FeNi after 3 MeV electron irradiation at 5 K [115, 159]. The peak of prime interest here is at 100 K. It is found to grow and subsequently to decay to about 25 % during annealing in stage I_{D+E}. It is attributed to a NiFe complex, tentatively to a <110> NiFe mixed dumbbell [158, 159], (Fig. 2.23). The growth of the peak occurs through the capture of migrating Fe self-interstitials at the beginning of stage I_{D+E} at Ni-atoms. The decay of the peak following its growth may be

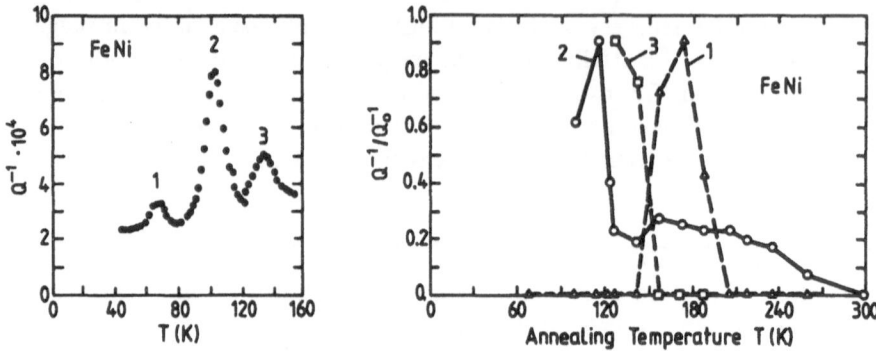

Fig. 5.17. Internal friction peaks observed in electron-irradiated FeNi (*left side*) and their respective annealing behavior. The concentration of the Ni atoms was 400 atppm, that of the Frenkel defects 250 atppm

attributed to the formation of larger Si complexes containing more than one trapped self-interstitial. This explanation suggests itself because of the observation that the amount of the decay increases monotonically with increasing self-interstitial concentration. The 100 K complexes remaining above stage *I* are very stable upon further annealing. Some are annihilated by vacancies migrating in stage *III* at about 210 K, but a small fraction apparently survives stage *III* and disappears shortly below 300 K.

The two other internal friction peaks occur at 45 K and 140 K. The 140 K complex is probably produced during the second half of stage I_{D+E}, parallel with the decay of 100 K internal friction peak. For this reason it seems natural to attribute the 140 K peak to trapped di-interstitials. Since the 45 K peak appears parallel to the decay of the 140 K peak, it may attributed to another higher order SI complex containing more than two self-interstitials or Ni-atoms.

The defect pattern given here is at variance with that developed on the basis of magnetic aftereffect studies of electron-irradiated Fe–Ni alloys [166]. Since internal friction and magnetic relaxation measurements are closely related techniques, one would expect similar results from both. Although a full discussion will be given in a forthcoming paper [159], the essential points may be presented here. There are in fact several corresponding processes in both spectra, but the two measurements differ in two essential points: the 100 K internal friction peak explained by the caging motion of a mixed <110> FeNi dumbbell does not have a corresponding counterpart in the magnetic aftereffect spectrum. Vice versa, a magnetic aftereffect at 107 K which was also interpreted in terms of caging of a mixed dumbbell [166] does not possess a counterpart in the internal friction spectra where it should appear at approximately 144 K. The activation energy of the latter process was found to be equal to the activation energy of the reorientation and simultaneous migration of the self-interstitials in stage I_{D+E}, i.e. 0.33 eV [165,166]. The activation energy of the 100 K internal friction peak, which is about 0.150 eV is considerably less than the activation energy for the self-interstitial migration. This lower activation energy, which one would expect for a mixed dumbbell, is therefore in favor of the assignment of the 100 K internal friction peak to the cage motion of a mixed dumbbell rather than the 107 K magnetic process. Another interpretation of the behavior of the self-interstitials has been given based upon studies of the recovery of the residual electrical resistivity [163, 164]. According to these, the migration of the FeNi mixed interstitial should occur in a resistivity recovery stage, II_{Ni}, located in the temperature regime from 130 K to 160 K. However, as the internal friction and magnetic aftereffect measurements show, the defect pattern in this temperature regime is quite complex and resistivity measurements cannot by themselves distinguish between the different defects.

6. Radiation-Induced Segregation

Radiation-induced segregation is one of the most prominent processes by which the microstructure of alloys suffers significant changes during irradiation. In its most simple form, enrichment or depletion of solute atoms at point defect sinks is observed in alloys subjected to high energy particle fluxes [126].

This alloy decomposition occurs by a preferential coupling of the defect fluxes — interstitial atoms and/or vacancies — with one of the alloy constituents, i.e. it is driven by kinetic processes rather than thermodynamic forces.

One example of such a coupling is the formation of the radiation-induced SI complexes in dilute alloys discussed in the previous section. If the binding energy is large enough, the majority of radiation-produced interstitial atoms will be trapped at solute atoms. If the migration energy of the SI complex is less than the dissociation energy, each migrating self-interstitial transports a solute atom. This is in contrast to the equipartitioning fluxes where the ratio of transported solute to host atom is given by the atomic concentration. Solute atom transport by an interstitial mechanism is thus a rapid and effective mechanism for preferential mass transport.

In a similar manner solute atoms may be preferentially transported via the formation of stable vacancy-solute atoms complexes.

More generally, preferential solute transport during irradiation will occur if there are any differences in the diffusion rates via vacancies or interstitials. This concept can also be used for concentration alloys. As shown by *Wiedersich* and *Lam* [127] the ratio of the steady-state concentration gradients of solute atoms, c_{SA}, and of vacancies, c_V, at a defect sink may be expressed as

$$\frac{\nabla c_{SA}}{\nabla c_V} \sim \left(\frac{D_{AV}}{D_{BV}} - \frac{D_{AI}}{D_{BI}} \right) \tag{6.1}$$

where D_{AV} is the partial diffusion coefficient for solute atoms via vacancies, D_{BV} for host atoms via vacancies, and D_{AI} and D_{BI} correspondingly for solute and host atom transport via interstitials. These coefficients may be written as [127]

$$D_{ij} = \left(\nu_{ij}^{\text{eff}} \cdot b_j^2 \cdot Z_j \right) / 6\Omega \tag{6.2}$$

where ν_{ij}^{eff} is an effective jump frequency, b_j is the jump distance, Z_j the coordination number, Ω the average atomic volume.

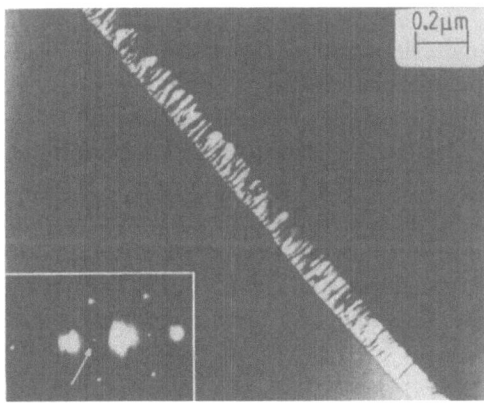

Fig. 6.1. TEM dark field image showing a layer of γ'–Ni$_3$Si grown on a grain boundary in a Ni-6 at% Si alloy, after 3.5 MeV Ni ion irradiation at 614 °C to 3.5 dpa. The insert shows the electron diffraction pattern exhibiting the superlattice spots due to the Ni$_3$Si phase, of which the one indicated by the pointer was used to take the image

The direction of the atomic mass transport is assumed to be opposite to the direction of the vacancy flux in (6.2).

Enrichment of solute atoms at defect sinks will always occur if the preference of self-interstitial for migration via solute atoms is larger than that of the vacancies.

One of the most striking and best documented examples of the phenomenon is the formation and growth of coherent Ni$_3$Si films on surfaces of ion-bombarded NiSi alloys [126, 128]. When NiSi alloys are irradiated point defect sinks such as dislocations, grain boundaries and surfaces become enriched with Si. At temperatures of about half the melting temperature and total doses of less than 1 dpa[1], the enrichment can exceed the solubility limit ($\sim 10\%$) and γ'-Ni$_3$Si will precipitate in an originally single phase alloy. An example of such radiation-induced segregation is shown in Fig. 6.1, which shows a roughly 100 nm thick layer of γ-Ni$_3$Si grown on a grain boundary.

The original alloy composition was 6 at%, the sample received irradiation with 3.5 MeV Ni$^+$ ions at 614 °C up to a total dose of 6 dpa.

The observed growth kinetics can be explained by a diffusion-controlled growth model in which the rate limiting step is the transport of fast migrating SI interstitial complexes to the surface [126]. Although definite experimental proof is still lacking, there is evidence to support the view that the segregation mechanism for Si is via interstitials as discussed in Chap. 5.

Radiation-induced alloy decomposition has been observed in a great number of cases. A compilation of data given in [126] demonstrates a rather striking correlation between the sign of radiation-induced segregation and the sign of the misfit of the solute atoms. Except for Al–Ge and Ni–Ge, depletion at defect sinks is always observed if $\delta_{SA} > 0$, whereas enrichment occurs for

---

[1]   dpa: displacement per atom, a measure for the total radiation-induced damage. It is the average number of displacements which an atom has experienced during the course of the irradiation.

undersized solute atoms, $\delta_{SA} < 0$. This observation is clearly in favor of the idea developed in Chap. 5, that undersized solute atoms are preferentially transferred into interstitial sites upon encountering a self-interstitial. On the other hand many of the experiments have been performed employing solute atom concentrations exceeding 1 at% where the validity of the simple complex formation model is hard to envisage. In addition, as shown above, the phenomena are rather insensitive to the properties of the transport process itself. In summary, although the decomposition process itself is well documented in many cases, the basic transport mechanism and the nature of the atomic species carrying this transport remains a subject for further research.

7. Conclusions

The investigation of radiation-induced interstitial atoms by means of mechanical relaxation techniques has provided valuable basic information about the structural and diffusional properties of such defects. The anisotropy factors $\pi^{(i)}$ provide the elastic dipole moments and the symmetry of the elastic dipole tensor which in turn can be used to test possible atomistic defect models. On the other hand, the relaxation strength has been used to follow the reactions of individual kinds of defects during isochronal annealing treatments. This was possible due to the fact that the relaxation lines of different kinds of defects distinguish themselves by their characteristic temperature locations. Thus, individual defects can be followed separately even in the presence of a larger ensemble of different species.

The variation of the temperature location of individual relaxation lines with the experimentally determined response times, e. g. vibrational period and elastic aftereffect times, yields the activation energy and the preexponential of the underlying diffusion process. In this sense, the case of the single self-interstitial in electron-irradiated Aluminum might be considered as a key-experiment and a highlight of the whole subject.

On the other hand, examples like electron-irradiated Copper or Molybdenum clearly reveal the limitations of the mechanical relaxation technique in that they show that orientationally undistinguishable defects remain undetected. In such cases — of course also in the former — full advantage can be taken from the complimentary nature of Huang diffuse x-ray scattering measurements. The latter ones provide the structural properties also in the case of isotropic or immobile defects, although they are insensitive to any effect involving the mobility of defects.

The interest in irradiation-induced interstitial atoms in alloys was originally related to phenomena in fission and fusion reactor materials. However, the significance of this subject extends into the more general field of non-equilibrium phenomena as they occur during ion implantation and ion beam processing of materials.

References

1   G. Leibfried, N Breuer: *Point Defects in Metals I* (Springer, Berlin, Heidelberg 1978)
2   P. Ehrhart, K.-H. Robrock, H. R. Schober: In *Physics of Radiation Effects in Crystals* (Modern Problems in Condensed Matter Science, Vol. 13), ed. by R. A. Johnson, A. N. Orlov (North Holland, Amsterdam 1986) p. 3
3   P. H. Dederichs, C. Lehmann, H. R. Schober, A. Scholz, R. Zeller: J. Nucl. Mater. **69–70**, 176 (1978)
4   K. W. Ingle, R. C. Perrin, H. R. Schober: J. Phys. F (Metal Phys.) **11**, 1161 (1981)
5   R. A. Johnson: Phys. Rev. A **134**, 1329 (1964)
6   R. Bullough, R. C. Perrin: Proc. R. Soc. London A **305**, 341 (1968)
7   A. Scholz, C. Lehmann: Phys. Rev. B **6**, 813 (1972)
8   R. Zeller, P. H. Dederichs: Z. Phys. B **25**, 139 (1976)
9   P. H. Dederichs, C. Lehmann, A. Scholz: Phys. Rev. Lett. **31**, 1130 (1973)
10 P. H. Dederichs: In *Fundamental Aspects of Radiation Damage in Metals* (USERDA, CONF-751006-P1, 1975) p. 187
11 P. H. Dederichs, C. Lehmann, A. Scholz: Z. Phys. B **20**, 155 (1975)
12 A. S. Nowick, B. S. Berry: *Anelastic Relaxation in Crystalline Solids* (Academic Press, New York, London 1972)
13 R. de Batist: *Internal Friction of Structural Defects in Crystalline Solids* (North Holland, Amsterdam, London; American Elsevier, New York 1972)
14 P. H. Dederichs: J. Phys. F (Metal Phys.) **3**, 471 (1973)
15 B. Alefeld, J. Völkl, G. Schaumann: Phys. Status Solidi **37**, 337 (1970)
16 J. Holder, A. V. Granato, L. E. Rehn: Phys. Rev. B **10**, 363 (1974)
17 V. Spiric: Berichte der Kernforschungsanlage Jülich, JÜL–1270 (1976)
18 H. Jacques: Berichte der Kernforschungsanlage Jülich, JÜL–1758 (1982)
19 K.-H. Robrock: Berichte der Kernforschungsanlage Jülich, JÜL–1088 (1974)
20 C. Zener: J. Appl. Phys. **18**, 1022 (1947)
21 J. v. Seggern: Private communication
22 S. Timoshenko: *Schwingungsprobleme der Technik* (Springer, Berlin 1932) p. 246
23 W. Schilling: J. Nucl. Mater. **69–70**, 465 (1978)
24 H. Ullmaier, W. Schilling: In *Physics of Modern Materials*, Vol. I (IAEA–SMR–46/105, 1980) p. 301
25 K.-H. Robrock, W. Schilling: J. Phys. F (Metal Phys.) **6**, 303 (1976)
26 L. E. Rehn, K.-H. Robrock: J. Phys. F (Metal Phys.) **7**, 1107 (1977); K.-H. Robrock, V. Spiric, L. E. Rehn: Radiat. Eff. **27**, 189 (1976)
27 L. E. Rehn, J. Holder, A. V. Granato, R. R. Coltman, F. W. Young, Jr.: Phys. Rev. B **10**, 349 (1974)
28 P. Ehrhart, W. Schilling: Phys. Rev. B **8**, 2604 (1973)
29 R. M. Nicklow, W. P. Crummett, J.M. Williams: Phys. Rev. B **20**, 5034 (1979)
30 F. W. Young, Jr.: J. Nucl. Mater. **69–70**, 310 (1978)
31 V. Spiric, K.-H. Robrock, L. E. Rehn: In *Fundamental Aspects of Radiation Damage in Metals* (USERDA, CONF–751006–P1, 1975) p. 240
32 V. Spiric, L. E. Rehn, K.-H. Robrock, W. Schilling: Phys. Rev. B **15**, 672 (1977)
33 M. Riggauer, W. Schilling, J. Völkl, H. Wenzl: Phys. Status Solidi **33**, 843 (1969)

34 K. Ehrensberger, K. Fisher, J. Kerscher, H. Wenzl: J. Phys. & Chem. Solids **31**, 1835 (1970)
35 R. A. Johnson: Phys. Rev. **152**, 629 (1966)
36 K.-H. Robrock, L. E. Rehn, V. Spiric, W. Schilling: Phys. Rev. B **15**, 680 (1977)
37 K.-H. Robrock, H. R. Schober: J. Phys. (France) **42**, C5–735 (1981)
38 H. R. Schober, R. Zeller: J. Nucl. Mater. **69–70**, 341 (1978)
39 G. De Keating-Hart, R. Cope, C. Minier, P. Moser: Berichte der Kernforschungsanlage Jülich, JÜL–Conf **2**, Vol. 1, 327 (1968)
40 J. R. Townsend, M. Schildcrout, C. Reft: Phys. Rev. B **10**, 14500 (1976)
41 L. E. Rehn, K.-H. Robrock, W. Schilling, V. Spiric: J. Nucl. Mater. **69–70**, 696 (1978)
42 P. Ehrhart: J. Nucl. Mater. **69–70**, 200 (1978)
43 H. Trinkaus: In *Fundamental Aspects of Radiation Damage in Metals* (USERDA, CONF–751006–P1, 1975) p. 254
44 P. Ehrhart, U. Schlagheck: J. Phys. F (Metal Phys.) **4**, 1575 and 1589 (1974)
45 K. Forsch, J. Hemmerich, H. Knöll, G. Lucki: Phys. Status Solidi a **23**, 223 (1974)
46 H. Knöll, U. Dedek, W. Schilling: J. Phys. F (Metal Phys.) **4**, 1095 (1974)
47 E. Segura, P. Ehrhart: Radiat. Eff. **42**, 233 (1979)
48 H. Schroeder, B. Stritzker: Radiat. Eff. **33**, 125 (1977)
49 P. Ehrhart: In *Fundamental Aspects of Radiation Damage in Metals* (USERDA, CONF–751006–P1, 1975) p. 302
50 H. Kugler, I. A. Schwirtlich, S. Takaki, U. Ziebart, H. Schultz: In *Yamada Conference V:* Point Defects and Defect Interactions in Metals, ed. by J. Takamura, M. Doyama, M. Kiritani (University of Tokyo Press, 1982)
51 J. Marangos: Published doctoral dissertation (TU Munich, 1980)
52 P. Ehrhart, H. Jacques, K.-H. Robrock: To be published
53 H. Jacques, K.-H. Robrock: J. Phys. (France) **42**, C5–723 (1981)
54 H. Jacques, K.-H. Robrock: In *Yamada Conference V:* Point Defects and Defect Interactions in Metals, ed. by J. Takamura, M. Doyama, M. Kiritani (University of Tokyo Press, 1982) p. 159
55 H. Mizubayashi, S. Okuda: Radiat. Eff. **33**, 221 (1977)
56 H. R. Schober: Privat communication
57 J. H. Evans: AERE–R 10 977, July 1983
58 K. Krishan: Radiat. Eff. **66**, 121 (1982)
59 H. Mizubayashi, S. Okuda: Radiat. Eff. **54**, 201 (1981)
60 H. Schultz: In *Yamada Conference V:* Point Defects and Defect Interactions in Metals, ed. by J. Takamura, M. Doyama, M. Kiritani (University of Tokyo Press, 1982) p. 183
61 R. R. Hasiguti: J. Phys. Soc. (Japan) **15**, 1707 (1960)
62 W. Schilling, C. Burger, K. Isebeck, H. Wenzl: In *Vacancies and Interstitials in Metals*, ed. by A. Seeger, D. Schumacher, W. Schilling, J. Diehl (North Holland, Amsterdam 1970) p. 255
63 R. W. Siegel: In *Yamada Conference V:* Point Defects and Defect Interactions in Metals, ed. by J. Takamura, M. Doyama, M. Kiritani (University of Tokyo Press, 1982) p. 533
64 S. Takaki, J. Fuss, H. Kugler, U. Dedek, H. Schultz: Radiat. Eff. **79**, 87 (1983)
65 H. R. Schober, R. P. Hatcher: Private communication
66 K.-H. Robrock, H. R. Schober: J. Phys. (France) C **5**, 735 (1981)
67 N. Q. Lam, N. V. Doan, V. Adda: J. Phys. F (Metal Phys.) **10**, 2359 (1980)
68 K. W. Ingle, R. C. Perrin, H. R. Schober: J. Phys. F (Metal Phys.) **11**, 1161 (1981)
69 H. R. Schober, R. Zeller: J. Nucl. Mater. **69–70**, 341 (1978)
70 H. R. Schober: J. Phys. F (Metal Phys.) **7**, 1127 (1977)
71 J. Bewerunge: Berichte der Kernforschungsanlage Jülich, JÜL–1615 (1979)
72 D. E. Becker, F. Dworschak, H. Wollenberger: Phys. Status Solidi b **47**, 171 (1971)
73 W. K. Warburton, D. Turnbull: In *Diffusion in Solids*, ed. by A. S. Nowick, J. J. Burton (Academic Press, New York 1975)

74 K. Schroeder, K. Dettmann: Z. Phys. B 22, 343 (1975)
75 H. W. King: J. Mater. Sci. 1, 79 (1966)
76 R. Lennartz, F. Dworschak, H. Wollenberger: J. Phys. F (Metal Phys.) 7, 2011 (1977)
77 H. Wollenberger: J. Nucl. Mater. 69-70, 362 (1978)
78 J. Selke: Diploma dissertation (RWTH Aachen, 1976)
79 F. Dworschak, C. Dimitrov, O. Dimitrov: J. Phys. F (Metal Phys.) 8, L153 (1978)
80 G. Kollers, H. Jacques, L. E. Rehn, K.-H. Robrock: J. Phys. (France) C 5, 729 (1981)
81 M. Deicher, R. Minde, E. Recknagel, Th. Wichert: V. International Conf. on Hyperfine Interactions, Berlin, 1980
82 C. P. Cannon, A. Sosin: Radiat. Eff. 25, 253 (1975)
83 R. Lennartz: Berichte der Kernforschungsanlage Jülich, JÜL-1400 (1977)
84 F. Dworschak, R. Lennartz, H. Wollenberger: J. Phys. F (Metal Phys.) 5, 400 (1975)
85 F. Dworschak, A. Kraut, K. Sonnenberg, H. Wollenberger: Radiat. Eff. 19, 119 (1973)
86 A. Kraut: Phys. Status Solidi b 44, 805 (1971)
87 F. Maury, A. Lucasson, P. Lucasson, J. LeHericy, P. Vajda, C. Dimitrov, O. Dimitrov: Radiat. Eff. 51, 57 (1980)
88 M. L. Swanson, L. M. Howe, A. F. Quenneville: Radiat. Eff. 28, 205 (1976)
89 H. Wollenberger: In *Fundamental Aspects of Radiation Damage in Metals* (USERDA, CONF-751006-P1, 1975) p. 582
90 M. L. Swanson, L. M. Howe, A. F. Quenneville: Can. J. Phys. 55, 1871 (1977)
91 A. Bartels, F. Dworschak, H.-P. Meurer, C. Abromeit, H. Wollenberger: J. Nucl. Mater. 83, 24 (1979)
92 F. Dworschak, Th. Monsau, H. Wollenberger : J. Phys. F (Metal Phys.) 6, 2207 (1976)
93 G. Kollers: Diploma dissertation (RWTH Aachen, 1981) unpublished
94 C. Dimitrov, F. Moreau, O. Dimitrov: J. Phys. F (Metal Phys.) 5, 385 (1975)
95 L. M. Swanson, L. M. Howe: J. Phys. F (Metal Phys.) 6, 1629 (1976)
96 L. M. Swanson, L. M. Howe, A. F. Quenneville: J. Nucl. Mater. 69-70, 372 (1978)
97 M. L. Swanson, M. L. Howe, A. F. Quenneville, P. Offermann, K. H. Eckert: J. Phys. F (Metal Phys.) 10, 599 (1980)
98 R. Butt, W. Semmler, R. Keitel: Phys. Lett. A 80, 29 (1980)
99 L. E. Rehn, K.-H. Robrock, H. Jacques: J. Phys. F (Metal Phys.) 8, 1835 (1978)
100 G. Vogl, W. Mansel: In *Fundamental Aspects of Radiation Damage in Metals* (USERDA, CONF-751006-P1, 1975) p. 349
101 W. Petry, G. Vogl, W. Mansel: Phys. Rev. Lett. 45, 1862 (1980)
102 W. Petry: Doctoral dissertation (HMI Berlin, 1980)
103 W. Mansel, H. Meyer, G. Vogl: Radiat. Eff. 35, 69 (1978)
104 C. C. Setser, K. L. Hultman, J. Holder, A. V. Granato: Bull. Am. Phys. Soc. 24, 242 (1979)
105 K. L. Hultman, J. Holder, A. V. Granato: Bull. Am. Phys. Soc. 24, 242 (1979)
106 D. L. Johnson, C. C. Setser, J. Holder, A. V. Granato: Bull. Am. Phys. Soc. 24, 242 (1979)
107 A. V. Granato: In *Yamada Conference V*: Point Defects and Defect Interactions in Metals, ed. by J. Takamura, M. Doyama, M. Kiritani (University of Tokyo Press, 1982) p. 67
108 P. R. Okamoto, L. E. Rehn: J. Nucl. Mater. 83, 2 (1979)
109 P. Moser: In *Internal Friction and Ultrasonic Attenuation in Crystalline Solids*, ed. by D. Lenz, K. Lücke, Vol. 1 (Springer, Berlin 1975) p. 239
110 J.-L. Oddou: CEA-R-3605 (1968)
111 R. S. Averback, P. Ehrhart: J. Phys. F (Metal Phys.) 14, 1347 (1984)
112 P. Ehrhart, R. S. Averback: J. Phys. F (Metal Phys.) 14, 1365 (1984)
113 P. R. Okamoto, L. E. Rehn, R. S. Averback, K.-H. Robrock, H. Wiedersich: In *Yamada Conference V*: Point Defects and Defect Interactions in Metals, ed. by J. Takamura, M. Doyama, M. Kiritani (University of Tokyo Press, 1982) p. 946
114 M. Weigert, A. Bartels, F. Dworschak: Mat. Sci. Forum 15-18, 1387 (1987)

115 C. Börner, H.-G. Bohn, K.-H. Robrock: Mat. Sci. Forum **15–18**, 617 (1987)

116 K. Nakata, K. Ikendii, H. Hirano, K. Furukawa, J. Takamura: In *Fundamental Aspects of Radiation Damage in Metals* (USERDA, CONF–751006–P1, 1975) p. 622

117 W. Heidrich: Berichte der Kernforschungsanlage Jülich, JÜL–1006 FF (1973)

118 E. Segura, P. Ehrhart: Radiat. Eff. **42**, 233 (1979)

119 W. Schilling, K. Sonnenberg: J. Phys. F (Metal Phys.) **3**, 322 (1973)

120 H.J. Dibbert, K. Sonnenberg, W. Schilling, U. Dedek: Radiat. Eff. **15**, 115 (1972)

121 K. Sonnenberg, W. Schilling, H.J. Dibbert, K. Mika, K. Schroeder: Radiat. Eff. **15**, 129 (1972)

122 W. Schilling, K. Sonnenberg, H.J. Dibbert: Radiat. Eff. **16**, 57 (1972)

123 K. Sonnenberg, W. Schilling, K. Mika, K. Dettmann: Radiat. Eff. **16**, 65 (1972)

124 C.Y. Wei, D.N. Seidman: Radiat. Eff. **32**, 229 (1977)

125 C.Y. Wei, D.N. Seidman: J. Nucl. Mater. **69–70**, 693 (1978)

126 L.E. Rehn, P.R. Okamoto: In *Phase Transformation During Irradiation*, ed. by F.V. Nolfi, Jr. (Applied Science Publishers, London, New York 1983) p. 247

127 H. Wiedersich, N.Q. Lam: In *Phase Transformation During Irradiation*, ed. by F.V. Nolfi, Jr. (Applied Science Publishers, London, New York 1983) p. 1

128 K.-H. Robrock, P.R. Okamoto: *Irradiation Behaviour of Metallic Materials for Fast Reactor Core Components*, ed. by J. Poirier, J.M. Dupony (CEA-DMECN Gif-sur-Yvette, France 1979) p. 57

129 H.R. Schober: To appear in physics B+C

130 W. Schilling, P. Ehrhart, K. Sonnenberg: In *Fundamental Aspects of Radiation Damage in Metals* (USERDA, CONF–751006–P1, 1975) p. 470

131 K.-H. Robrock: In *Phase Transformation During Irradiation*, ed. by F.V. Nolfi, Jr. (Applied Science Publishers, London, New York 1983) p. 115

132 H.G. Haubold, D. Martinsen: J. Nucl. Mater. **69–70**, 644 (1978)

133 H.R. Schober: J. Nucl. Mater. **126**, 220 (1984)

134 R. Urban, P Ehrhart, W. Schilling, H.R. Schober: Mat. Sci. Forum **15–18**, 243 (1987)

135 J.H. Evans: J. Nucl. Mater. **132**, 147 (1985); see also: Mat. Sci. Forum **15–18**, 869 (1987)

136 P. Lucasson, F. Maury, A. Lucasson: Radiat. Eff. Lett. **85**, 219 (1985)

137 N.Q. Lam, N.V. Doan, L. Dagens, Y. Adda: J. Phys. F (Metal Phys.) **11**, 2231 (1981)

138 N.Q. Lam, L. Dagens, N.V. Doan: J. Phys. F (Metal Phys.) **13**, 1369 (1983)

139 F. Maury: Radiat. Eff. **60**, 181 (1982)

140 H. Wollenberger, C. Abromeit, F. Dworschak: Radiat. Eff. **70**, 255 (1983)

141 C.C. Setser, K.L. Hultman, J. Holder, A.V. Granato: Phys. Rev. B **32**, 1453 (1985)

142 P.W. Wallace, W.L. Hultman, J. Holder, A.V. Granato: J. Phys. (France) C **10**, 59 (1985)

143 A. Bartels, F. Dworschak, M. Weigert: J. Nucl. Mater. **173**, 130 (1986)

144 C.P. Flynn, A.M. Stoneham: Phys. Rev. B **1**, 3966 (1970)

145 H. Schultz: Mat. Sci. Forum **15–18**, 727 (1987)

146 W. Frank: Mat. Sci. Forum **15–18**, 875 (1987)

147 G. Sulpice, C. Minier, P. Moser, H. Bilger: J. Phys. (France) **29**, 253 (1968)

148 W. Dander, H.E. Schaefer: Phys. Status Solidi b **80**, 173 (1977)

149 H.E. Schaefer, W. Dander: Phys. Status Solidi b **78**, 139 (1976)

150 R. Pichon, E. Bisogni, P. Moser: Radiat. Eff. **20**, 159 (1973)

151 R. Pichon, E. Bisogni, P. Moser: Radiat. Eff. **22**, 173 (1974)

152 A. Lucasson, V. Loreaux, F. Maury, P. Lucasson: J. Phys. F **14**, 1379 (1984)

153 F. Maury, P. Lucasson, A. Lucasson, P. Vajda: Radiat. Eff. **82**, 141 (1984)

154 F. Maury: J. Phys. F **14**, 1395 (1984)

155 F. Maury, A. Lucasson, P. Vajda, J. LeHericy, A. Mathieu, C. Dimitrov, P. Lucasson: Radiat. Eff. **55**, 187 (1981)

156 M. Kobiyama, S. Takamura: Phys. Status Solidi a **90**, 253 (1985)

157 S. Takamura, M. Kobiyama: Phys. Status Solidi a **90**, 269 (1985)

158 K.-H. Robrock: Mat. Sci. Forum **15–18**, 537 (1987)

159 C. Börner, K.-H. Robrock: To be published

160 K.-H. Robrock: In *Dimensional stability and mechanical behaviour of irradiated metals and alloys*, British Nuclear Energy Society, London, Vol. 1, 105 (1983)

161 H. R. Schober, A. M. Stoneham: Phys. Rev. B **26**, 1819 (1982)

162 H. Wollenberger: In R. W. Cahn, P. Haasen: *Physical Metallurgy*, Part II (North Holland, Amsterdam, Oxford, New York, Tokyo 1983) p. 1139

163 F. Maury, A. Lucasson, P. Lucasson, Y. Loreaux, P. Moser: J. Phys. F (Metal Phys.) **16**, 523 (1986)

164 F. Maury, A. Lucasson, P. Lucasson, P. Moser, Y. Loreaux: J. Phys. F (Metal Phys.) **15**, 1465 (1985)

165 H. J. Blythe, F. Walz, H. Kronmüller: Phys. Status Solidi a **69**, 237 (1982)

166 H. J. Blythe, F. Walz, H. Kronmüller: Phys. Status Solidi a **81**, 227 (1984)

167 P. Vigier, V. Hivert, P. Moser, E. Bonjour: C.R. Acad. Sci. Paris **260**, 3359 (1965)

168 C. Zener: *Elasticity and Anelasticity of Metals* (The University of Chicago Press, Chicago, London 1948)

Subject Index

Springer Tracts in Modern Physics

Volume 91

K. Heinz, K. Müller, T. Engel, K. H. Rieder

Structural Studies of Surfaces

1982. 180 pp. ISBN 3-540-10964-1

"This is and excellent, up-to-date, authoritative and balanced account of two particular approaches to the elucidation of surface structure. The book, which is well produced, is suitable for those already involved or embarking on research in the field of surface science where the emphasis is mainly on the application of modern physical techniques."
Journal of the Chemical Society, Farady Transactions

"This is an invaluable reference guide for anyone working in the field of atomic and molecular beam scattering at surfaces (experimental and theoretical)." *J. Am. Chem. Soc.*

"In summary, this is a worthy addition to the Springer Tracts and can be recommended enthusiastically."
Journal de Physique

Volume 94

V. M. Kenkre, P. Reinecker

Exciton Dynamics

1982. 226 pp. ISBN 3-540-11318-5

"... an important and useful introduction for theorists and experimentalists alike.
... I regard this as a well-written useful book in the area of theoretical-condensed matter science."
Applied Optics

Springer-Verlag Berlin
Heidelberg New York London
Paris Tokyo Hong Kong

Springer Tracts in Modern Physics

Volume 103

H. Coufal, E. Lüscher, H. Micklitz, R. E. Norberg

Rare Gas Solids

1984. IX, 99 pp. 40 figs. ISBN 3-540-13272-4

"... Because of its detailed description of the different experimental techniques... this book may be recommended not only to the specialist but also as an introduction..."

Crystal Research and Technology

Volume 104

I. Pockrand

Surface Enhanced Raman Vibrational Studies at Solid/Gas Interfaces

1984. IX, 164 pp. 60 figs. ISBN 3-540-13416-6

"... The book should thus be useful to surface scientists interested in the vibrational spectroscopy of adsorbates as well as to others requiring a review of this important new field."

Philosophical Magazine B

Volume 110

A. Stahl, I. Balslev

Electrodynamics of the Semiconductor Band Edge

1987. IX, 215 pp. 42 figs. ISBN 3-540-16953-9

"... The book... should be useful to theoreticians interested in optical effects in semiconductors. Since its initial chapters are written in a self-contained, pedagogical manner, graduate students may also find it useful..."

The Australian Physicist

Springer-Verlag Berlin
Heidelberg New York London
Paris Tokyo Hong Kong